Contemporary Issues in Economic Development, Africa

Contemporary Issues in Economic Development, Africa

Prof. V.A. Adeyeye
Director, Economics Policy Unit
Nigeria Institute of Social & Economic Research (NISER)
Ibadan, Oyo State, Nigeria

and

Dr. Olayiwola Olubodun Olaniyi
Nigeria Institute of Social & Economic Research (NISER)
Ibadan, Oyo State, Nigeria

NEW INDIA PUBLISHING AGENCY
New Delhi – 110 034

NEW INDIA PUBLISHING AGENCY
101, Vikas Surya Plaza, CU Block, LSC Market
Pitam Pura, New Delhi 110 034, India
Phone: + 91 (11)27 34 17 17 Fax: + 91(11) 27 34 16 16
Email: info@nipabooks.com
Web: www.nipabooks.com

Feedback at feedbacks@nipabooks.com

ISBN: 978-93-85516-35-1

Composed, Designed and Printed in India

DEDICATION

I dedicate this book to God Almighty and entire family and friends for their contribution towards the success of the book.

Preface

The book has been compiled bearing in mind a variety of needs from business peoples that practice small scale business and the language and style of the book is consciously made simple so as to effectively cater for the multiplicity of interest groups highlighted. I must add here that the bulk of materials in this book had come in the main from a compilation of field surveys and Focus Group discussions, Interviews, questionnaire and literature reviews on the subject matters. The field of economics is wide and characterized by tortuous terrain. Although there are several texts and publications in economics, books which focus on the adaptation of economics principles to the entrepreneurship system are not very common. Traders, government, students of economics and business management therefore face the trauma of adapting principles of entrepreneurship to economic systems. These gap which this present book along with its precursors attempt to bridge. In order to fulfill this objective, the structure of the book is design to present an expose of entrepreneurship principles and small scale business in Nigeria.

Authors

Contents

Chapter 1

Small Scale Enterprises

1.1. Definition of Small-scale Enterprises

According to Shonekan's keynote address at the Faculty of Business Administration Conference (1992), captured with the theme: "Management of Small and Medium Scale Enterprises in Nigeria." He said,

"While it is difficult to have one universal definition of SMEs since there are usually many non-constant factors and variables, the spin-off industries are usually small or medium scale. In practical terms, SMEs constitute the bedrock of our national economy. They are our industrial feedstock, the engine of production for the bigger industries. The thriving economies of the world have a record of carefully structured SMEs. In Japan for instance, the bulk of manufacturing is at the small and medium scale levels. This situation also obtains in the 'Asian Tigers' (i.e. Malaysia, Taiwan, South Korea, Singapore, etc)."

The strength of an enterprise lies in its ability to exploit its environment to acquire scarce and valued resources. Thus, effectiveness entails production, cost, turnover, quality of output, profitability, efficiency, etc. (Katz and Kahn, 1966).Enterprise is considered here, as a business concern initiated by the entrepreneur who employs in his service other people for the primary purpose of achieving stated or implicit goals for which such business is established. The word "*Entrepreneur*" is also used here to mean the innovating individual who initiates and nurtures to growth an on-going business organization where none existed before.

In Nigeria, different government agencies have given their own definition of what a small-scale business is. Under the government's small-scale development programmes any manufacturing, processing and service industry with a capital investment not exceeding ₦ 150,000, in machinery and equipment alone is to be treated as a small-scale industry, Ejiofor (1984). The 1989 industrial policy of

Nigeria defined small scale industries as those with total investment of between ₦100,000.00 and ₦ 2 million, exclusive of land but including capital. While micro and cottage industries are defined as those with investment not exceeding ₦ 100, 000.00, exclusive of land but including working capital.

The third National Development Plan (1975-1980) defined small-scale business thus "manufacturing establishments employing less than ten people or whose investment in machinery and capital, equipment do not exceed ₦ 600,000" Mbanefor (1979).The CBN in its own interpretation insists that any business whose turnover is above ₦ 500,000 does not qualify to be described as a small-scale business. Most of the definitions of small-scale business appear to be governed by the interest of the perseverance purpose of definition. For the purpose of this study, a small-scale business may be defined as the one which is owned and managed by one or two persons, is family influenced in decision making, has undifferentiated organizational structure, has a relatively small share of the market and employees less than 25 people.

1.2. Problems of Small-scale Business

Problems or failure had been an inherent phenomenon in the business world. At any one time, an entrepreneur has to battle against problems or failure. Many projects wind up or are abandoned completely at a premature stage in its life, it is a common phenomenon among small business firms. The root of the problem lies in the managerial capacity of these firms and the orientation of the owners of the business. Many Nigerians in business have lazy ideas. Articulate and purposeful planning is not topmost on their priority list. Feasibility studies are not mostly conducted to determine the viability or otherwise of a proposed venture in developing countries such as India and Korea, where definite skilled and well thought out plans for entrepreneurship is the watch word.

1.2.1. Classification of SMEs

Available literatures on the subject matter of SMEs show a common consensus on the criteria used for identifying SMEs. They include the number of persons employed, the value of investment, annual turnover, or a combination of these. While there is a consensus regarding the available criteria, there is however no hard and fast rule for classifying SMEs. What, therefore, constitutes a micro, small and medium-scale enterprise varies across space and time. For instance, what may be regarded as a micro or small-scale enterprise in an industrialized economy may be regarded as a medium or large-scale firm in a developing country. Similarly, within the same country, classification of SMEs varies with time and among agencies. Such variation over time is attributable to such factors as movement in the general price level and technological advancement.

According to IFC (2003), the World Bank Group's SME Department classified micro and SMEs based on the following criteria:

Micro-enterprise

Employees:	10 or less
Total Assets:	US$ 100,000 or less
Total Annual Sales:	US$ 100,000 or less

Small enterprise

Employees:	Between 10 and 50
Total Assets:	US$ 100,000 to $3 million
Total Annual Sales	US$ 100,000 to $3 million

Medium enterprise

Employees:	50 and 300
Total Assets:	US$ 3 million to $5 million
Total Annual Sales	US$ 3 million to $5 million

In the same vein, the National Council on Industry, in its 31st meeting held in Makurdi in 2001 classified SMEs as follows:

- *Micro/cottage Industry*: An industry with total capital employed of not more than ₦ 1.50 million, including working-capital but excluding cost of land, and/or a labour size of not more than 10 workers.
- *Small-Scale Industry*: An industry with total capital employed of over ₦ 1.5 million but not more than ₦ 50 million, including working capital but excluding cost of land, and/or a labour size of 11-100 workers.
- *Medium-Scale Industry*: An industry with total capital employed of over ₦ 50 million but not more than ₦ 200 million, including working capital but excluding cost of land and/or a labour size of over 300 workers.
- *Large-Scale Industry*: An industry with total capital employed of over ₦ 200 million, including working capital but excluding cost of land, and/or labour size of over 300 workers.

Nigeria, like all developing countries, is increasingly looking to the Small and Medium Enterprises (SME) sector to absorb its growing workforce. As part of the reform process in the public service, public institutions are to be right sized. The staffs that are disengaged as a result of this policy are expected to move into the SME sector as new entrepreneurs. However, the small enterprise sector is very heterogeneous. The small enterprises vary in size, in type of business they are engaged in, in the industry sector they are operating in, in the products produced or services rendered, in the processes and level of technology used and in the business environment that they are located. Similarly, a number of needs of entrepreneurs are common worldwide. These include the need to generate new business ideas, management know-how, marketing, access to finance and information on a whole range of issues. In the process of discussing the principles of entrepreneurship and small-scale business management, a number of these need areas will be focussed upon. In order to properly guide our discussions, it is important to state clearly what constitute a small-scale enterprise. The terms small, medium and large scale are relative and their definitions are as varied as the country and sector. There is no unique or universally accepted definition of the terms and the definitions even change with time depending on the level of development of the country. There are, however, some variables, namely capital asset or turnover (gross output) and number of paid employees which are often used in determining the scale.In Nigeria, the Federal Ministry of Industries (1993) defines industries in terms of asset base as follows:

i. *Cottage industry or micro enterprise:* An enterprise whose total project cost, excluding cost of land but including working capital, is not more than US $ 50,000.

ii. *Small-scale Enterprise:* An enterprise whose total project cost, excluding cost of land but including working capital, does not exceed US $ 500,000. In terms of number employed, a cottage industry or micro enterprise employs no more than 10 people while a small-scale enterprise employs between 11 and 50 people.

The entire series of activities that must be covered from the business idea conception stage to the stage where the business is operationalised is referred to as the Business Cycle. One of the simplest models that show the business sequence is the Baum's cycle which is shown in Figure 1.

Conception/Identification Stage: At this stage, a number of business ideas are generated and the most feasible is selected for proper planning.

Preparation Stage: Preliminary planning is done and the project document which in the case of an industrial project, a feasibility report is prepared.

Appraisal Stage: The project document at this stage is subjected to thorough analysis in order to determine the feasibility of the business ideas. At the end of the appraisal, a Go/No go decision must be taken. These three stages provide the entrepreneur awareness and pre-start up preparation. If a Go decision is

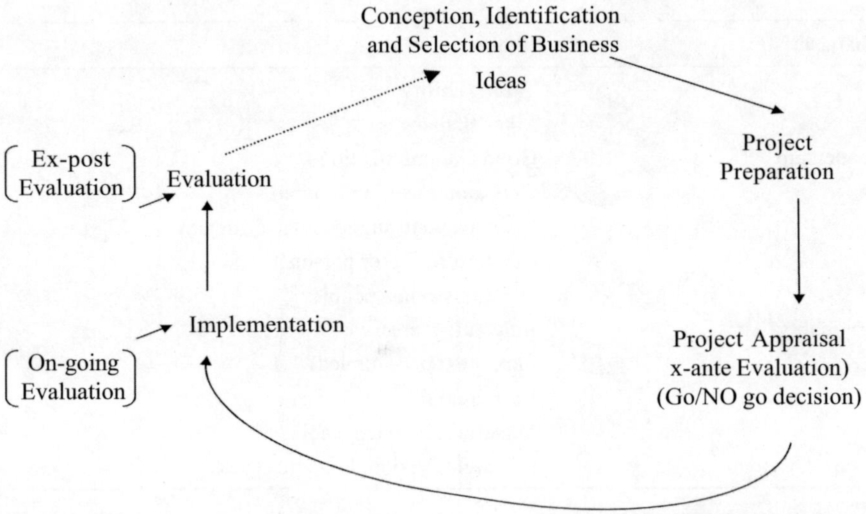

Fig. 1: The Baum's Cycle of Business Sequence

taken, then the entrepreneur is ready to start the business. Having operationalised the business for some years, some post mortem evaluation has to be done and in the process new business ideas may emerge for further investigations. The entrepreneur is the most important requirement for the success of a small business. The entrepreneurs are those who create new business ventures, assume the personal and business risks associated with the new business and continue to actively manage its operations. The entrepreneurs perform two distinct but interrelated functions, entrepreneurship and management. Entrepreneurship is the quality which enables people to start a new business or vigorously and innovatively expand an existing one. An entrepreneur generally exhibits some of the characteristics shown in Table 1.

Table 1: Profile of an Entrepreneur

Characteristics	Traits
Self-Confidence	Belief in own ability
	Independence
	Optimism
Strong Will Power	Persistence and Perseverance
	Determination
Task/Result Oriented	Achievement oriented
	Hard work, drive, energy
	Initiative
Risk Taker	Risk assessment and judicious risk

Contd....

Characteristics	Traits
	Taking ability
	Likes challenges
Leadership	Good Communication
	Gets along well with others
	Responsive to suggestions, criticism
	Concern for other person
	Develops other people
Originality	Innovative, creative
	Flexible (open-minded)
	Resourceful
	Versatile, Knowledgeable
Future-Oriented	Foresight, Vision, Perceptiveness

Source: Arturo Tolentino (1997): "Training and Development of Entrepreneur-Managers of Small Enterprises", Enterprise & Management Development, ILO.

The entrepreneurs are able to scan their environment to see business opportunities. They link their small business to the environment through formal and informal network of relationships with family members, friends and other business people. These enable them to get necessary information on new technologies, best practices, market trend, support and resources essential to the start-up and running of their small business. In order to perform its entrepreneurial functions effectively, entrepreneurs need to develop some competencies either through practice and experience or proper training. These competencies are shown in Table 2.

Table 2: Competencies Required to Perform Entrepreneurial Functions

Competencies	Characteristics
Perceptiveness	The ability to be alert for and identify new business opportunities.
Analytical Ability	The ability to analyse an opportunity or a problem situation, to be able to look at the situation holistically and to break it down into components allowing the analysis of cause and effects and other relationships.
Decision-Making	Identification and evaluation of alternatives and selection of what appears to be the best solution or course of action to take.
Learning Ability	The ability to derive lessons and insights from experiences and to apply these learning to other situation.
Initiative	The ability to be self-driven and motivated and to keep on going without waiting to be told to do so by others.

Contd....

Competencies	Characteristics
Empathy	The ability to identify oneself mentally with customers, clients, suppliers, competitors, service providers and other stakeholders in the business environment thus comprehending better their needs, expectations, apprehensions and requirements.
Persuasion	The ability to communicate with others (customers, suppliers, partners, workers, government officials, etc) make them understand and convince them on a particular proposition.
Negotiating Ability	The ability to discuss the various sides of an issue or proposal and strike agreements that are mutually advantageous and acceptable to all parties involved.
Networking	The ability to establish linkages with other business persons and other stakeholders for mutual learning, collaborative undertakings and other joint activities aimed at achieving common objectives.

Source: Ibid.

1.2.2. Ways of owning a small-scale business

A new business can be owned on any of the three main ways:

- Buying over an existing business;
- Inheriting a business; or
- Launching a new business.

Buying an existing business venture is quite advantageous in some respects. Firstly, the profitability of the business can ascertained from the records of the business. The existence of established customers will further eliminate uncertainties concerning the business, moreover, the exiting business is similar to buying an existing one. The only differences lies on the fact that one need not to pay a price for the acquisition, Keith A.L., and Guberlling, C.E. (1975). However, when *inheriting business* it is advisable of the heir to establish the worth of the business via an independence and objective audit. *Starting a new business* from the scratch carries with it a heavy risk and burden that it does in the case of the other two ways of owning a business. Adequate preparations resourceful, capable and dynamic management can transform a new business into a giant concern. This is because the owner selects his location, products and service merchandise, equipment, workers, suppliers and bankers. He is also able to originate his own policies and practices and attract customers.

1.2.3. Forms of SME ownership

The main types of ownership are the sole proprietorship, the partnership and the corporation or the limited liability company.

Sole proprietorship

It is the oldest, simplest and commonest form of business to bring into fruition. It is also the easiest to dissolve; it represents the will of a single individual to put his business ideas into practice. This form of business ownership is the most prominent in the Nigerian society for it has no requirements to meet.

No formal document need be drawn up and no organizational fee or taxes need be registration of the business name where it differs from the surname of the proprietor.

The proprietor takes all the risks of loss or failure of the business, but at the same time, he recovers the whole profits should the business succeed as a 'going concern." (Olowoyeye, 1986).

Partnership

It has been described as the relationship, which subsists between persons carrying businesses in common with a view of making profit. The partnership act of 1890 sets down in general term the circumstances in which persons are regarded as acting partnership and the legal implication of such an agreement which may be in writing, deals with the amount invested on the business by each partner, the rules for the sharing of profits and losses, the addition of new members and dissolution. The partnership agreements contain such information as the name of partners, the name of the commencement of the business and a provision for the settlement of disputes. Each partner's acts as a general agent for the partnership accept the partners in the agreement have decided to limit individual liabilities and responsibilities.The partnership exists only by the will of and during the lifetime of a partner. The death, insanity, bankruptcy and withdrawal of a member or the admission of a new member automatically requires the dissolution of the old partnership and the formation of a new one (Cole, 1984).

1.2.4. Selecting the mode of owning a SME and form of ownership

Determining the mode of owning a business will depend on the circumstances and the intent of the potential business owner. The decision to be made here is strictly between buying an existing business and floating a new one from the beginning. Final decision will depend on the cost benefit analysis. A person who is in a position to inherit a business can either take over the operation of the business or sell it. The two different forms of SME ownership have the advantages and disadvantages, a competent entrepreneur should weigh these before deciding on the form of ownership he will take. It should be made clear here that there is no step in business formation that is more important than the form of the ownership of the business affects not only the sharing of taken by the business, its ability to raise funds and other resources.

The efficiency of various forms of SME ownership on dealings with fundamental business problems, organization, the ease and cost of formation, operation and management, be highlighted in the following discussion, Harold Kootz Cyril O'DonnelWeirch (1980).

1.2.5. Operation, ownership and management

The operation of one-man business (i.e., sole proprietorship) has some advantages over partnership. The fact that the business cannot be divulged from the person of the owner, spurs him to a greater and better performance. In managing the business, greater flexibility is available to sole proprietor in making decisions. The control of the business is exclusive in his hands. Disadvantages generally include lack of continuity on most cases at the demise of the owner; the nature of the business ownership does not encourage the pooling of sources from different people.

The partnership like sole-proprietorship benefits from strong motivation of partners from the merging of abilities to facilitate efficient management. But the disadvantage of partnership includes the loss of security or strategies and plans and the diminished flexibility in the decision making because of the split in authority.

1.2.6. Factors determining small and medium scale entrepreneurship choice

There are several spheres in business to which forecasting can be directed. For the purpose of this study, the focus will be on the market, the economic environment, socio-cultural environment, political environment and legal environment.

Market factor

Many small forms of business fail because they have not assessed properly and articulate the demand pattern. Many of these businesses are established based on figments of imagination that market exists for goods and services.

A wise entrepreneur will systemically scan the market to see if the level of demand for proposed goods and services would justify his efforts. If the point to the contrary, the decision to launch the enterprises should be dropped, Appelel (1981).

Economic factor

The dynamic environment in which the business operates is full of threats and opportunities. Efforts should be made to predict the future trend and adjust on anticipation of new development. One should not fold his arms and wait, for it may be late to catch up with changes which the new development brings with it.

Political factor

In Nigeria and other developing countries, the government plays a central role in economic planning. Therefore, a great deal of information may obtain and form periodic and annual development plans, budgets, and other government services.

The rate of turnover of political leaders comes with different ideas about how the country should be administered. Where the turnover is high, projects with shorter period are recommended.

Social-Cultural factor

This can be defined as the aggregate of attitudes, beliefs and customs of the people in a given society. An entrepreneur should study the socio-cultural environment way well before reaching the decision to invest. If the people in a given area are predominantly Muslims, establishing a hotel where wine, spirits and other alcoholic beverages is to be sold, will not stand the test of time, Adesanya (1972).

Technological factor

For a business to compete favourably with its competitors, it has to be abreast with latest technology in its area of operation. Technology impact on business in terms of efficiency and effectiveness.

1.2.7. Policy formulation in SMEs

'Micro, small and medium - sized enterprises (SMEs) are the engine of the European economy. They are an essential source of jobs, create entrepreneurial spirit and innovation in the EU and are thus crucial for fostering competitiveness and employment. (Günter Verheugen, Member of European Commission 2005. The new SME definition, User guide and model declaration) "The category of micro, small and medium - sized enterprises (SMEs) is made up of enterprises which employ fewer than 250 persons and which have an annual turnover not exceeding 50 million euro, and/or an annual balancesheet total not exceeding 43 million euro." (Extract of Article 2 of the Annex of Recommendation 2003/361/EC).

Defining SMEs is challenging, because there is no single agreed definition of an SME and numerous are applied among OECD (Organization for Economic

Co-operation and Development) countries, and employee numbers are not the sole defining criterion. SME are usually independent non subsidiary firms which employ certain number of employees. In the European union the most frequent upper limit designation an SME is 250 employees, but some countries can set the limit to 200, while over the sea The United States consider SMEs to include firms with fewer than 500 employees. (OECD 2005) EU commission changed the definition of SMEs in 2005 from the previous one that was adapted in 1996. The changes were mostly done because of general economic developments since 1996, and a growing awareness to the specific barriers confronting SMEs. The new SME definition is done for all business categories and also takes better account of different types of relationships between enterprises. The new definition also ensures that different support measures are awarded only to the ones that need them.

Changes are also done because the businesses are now days more innovative and technology plays a crucial role as well, (EU Commission). It is important to define SMEs, in implementing efficient measures and programs to support SMEs in every possible way. SMEs play a big role in the creation of jobs and a country's employment rate. The most evident public benefit of small business growth is the contribution made by SMEs to employment. A large number of studies carried out in various countries have concluded that small business plays major role in job creation (Hamilton, 2007, 297).

A key characteristic of an SME is flexibility, the ability to adapt to changing environment (Levy & Powell 2005). These days there is lots of pressure for firms to be flexible in developing and developed countries. Companies are facing external pressures like a transforming market and a high speed technological change.

Borrowing a leaf from Large Scale Enterprises, Small and Medium Scale enterpreneurs should embark on strategic planning. Strategic planning areas in a business be it small or big, include the formulation of policies not only on personnel, marketing, production, finance and insurance but also on profitability. Policies emanate from plans. The area general statements are designed to guide the activities of business operations towards the attainment of the corporate goal. It provides the structure on which future activities can be based.

Financial and investment policies

In the course as new business, the financial and investment policies should be given pride of place. They are very important because of their profound effects on profitability and survival of the new business. Considerations should be given to the structure of the firm.

Capital structure, sometimes known as financial plan points to the composition of long-term sources of funds, such as debentures, long-term debts, preference shares, capital and equity shares, capital including reserves and surplus.

Some businesses do not plan their capital structure; it develops somewhere along the line with the decisions of the financial manager. Such company can start with a shabby arrangement in the short-term, but ultimately, they will encounter difficulties in raising funds to finance their activities.

Therefore, a SME should plan its capital structure to maximize the use of funds. A retained earnings is one significant way of financing a business on the long run; no serious cost is involved in getting this fund. On the other hand, investors deserve at least fair dividend over a period to justify their investments.

A balance has to be stricken between a long-term financing decision and wealth maximization decision, Schmoth (1993).

Personnel policies

The central theory in the firm's personnel policy is to find the right man for the right job so that his talent may be polished and the objective of the organization is realized.

Due recognition must be given to on-the-job-training. While it may be difficult to train people to spot economic opportunities, it is possible to train them, and assess such opportunities.

Similarly, certain managerial skills are trainable, but without them, new SMEs might not survive because of their inability to overcome initial management difficulties.

Marketing policies

Marketing policies, which will give a clear answer to the following questions should be formulated:

a. What are the various channel of distribution?

b. What should the pricing and credit policies be?

c. What will be the branding and packaging policies?

Sound policies, which can give answers to the questions above, will enhance the success of the business?

Management policies

In the first place, organizational structure should be drawn up to indicate the various positions in the enterprise and the flow of authority be it partnership or sole proprietorship that has employees. Furthermore, administrative rules and regulations should be made abinitio. This will go a long way to eliminate trumped up charges or offence against employees by their peers and employers, Keith and Gubelling (1975).

Management controls events plans. Control is essential managerial function, which points out weakness and errors in order to rectify them and prevent reoccurrence. Feed forward for earns what the future holds in stock for the business and making arrangement to ensure the actual performance against standard. This is not necessarily true in all cases. Feed forward who is drawn from forecasting can help in controlling business activities, Billies (1974).

Insurance policies

Engaging in a business venture is risky. No matter how prepared one is, success is not fully guaranteed, in the business world some of the risks inherited in running small-scale business ventures are fire, burglary, accident, death etc.

Insurance, which is one of the most important measures for protecting business, may mean the difference between continued survival and bankruptcy. An astute business man will shift some of these risks to another person or institution's such as insurance companies where there is opportunity to do so. This he can do by paying insurance premiums. For compensations which will be paid in the event of a specified unfortunate occurrence.

It is important that the small-scale businessman makes regular risk survey of his business, he should be able to recognize major risk situations and determine which risk can be assumed or transformed. The small-scale businessman should be warned of the chances of under-insurance. Having two little coverage of any type of insurance is dangerous even through the premium cost is reduced by reduction of the coverage. It is also important to wary in the choice of insurers.

The ability to settle claims promptly and adequately should be the major determinants of the choice of insurers. The insurance company's annual reports and statement of account could be a valuable source of information here, Gesternberg (1959).

1.3 The SME Option: Its Problems

The problems facing SMEs in Nigeria are not insurmountable since our country is well placed to provide the resources to solve these problems. In the main, the problems range from inadequate financing, lack of accountability to lack of infrastructure.

Most small businesses are a one-man proprietorship and the proprietor, through having to give his attention of various managerial functions in his business soon loses full grip. This is why it makes for efficient management to spread the responsibilities within the business or seek the external assistance of a Management Consultant. The SMEs also invariably require a high technical input and the cost of this technical expertise is usually high.

The issue of inadequate financing usually due to inadequate proprietorship equity participation. A strong capital base and financial outlook are required for a successful business. It therefore makes for sound financial management to seek for more business partners who could bring in more funds to finance the activities of the enterprise. However, small scale entrepreneurs are usually reluctant to bring in partners, even at the risk of under-capitalization. The underfunding of SMEs has led government to formulate measures aimed at assisting this sector. Loan packages of the World Bank, National Directorate of Employment (NDE) and the National Economic Reconstruction Fund (NERFUND) are among such assistance schemes.

The one-man proprietorship syndrome of SMEs is not only bad for their financial profiles but can discourage accountability which is the foundation of a successful business. Accountability ensures adequate control over the business and makes it possible to work out appropriate financial strategies.

The SMEs are not shielded or immuned to the lack of infrastructure, which is the bane of all business in Nigeria. On the contrary, SME's groan more under the weight of overhead costs of supplying their own boreholes, generating plant, high cost of telephone and electricity. Needless to say, no business can survive without these facilities. I believe that government should improve the infrastructural base of the economy such that the cost of development of SMEs would be reduced.

We all know the complexity of frustration involved in sourcing for foreign exchange to take care of import needs of a business. While I accept that most businesses, including SMEs, cannot do without foreign exchange altogether to get their business moving, it is my view that SMEs should adopt a policy of sourcing for inputs locally to reduce their costs. Having said that, in a situation where some level of hard currency will be required by SSMEs I am of the view that the Government should make foreign exchange affordable to SMEs and allow the naira to float to find its true value, we would not realize any advantage from the devaluation of CFRA franc.

Problems or failure had been an inherent phenomenon in the business world. At any one time, an entrepreneur has to battle against problems or failure. Many projects wind up or are abandoned completely at a premature stage in its life, it is a common phenomenon among small business firms. The root of the problem lies in the managerial capacity of these firms and the orientations of the owners of the business. Many Nigerians in business have lazy ideas. Articulate and purposeful planning is not topmost on their priority list. Feasibility studies are not mostly conducted to determine the viability or otherwise of a proposed venture in developing countries such as India and Korea, where definite skilled and well thought out plans for entrepreneurship is the watch word.

(a) Market factor

Many small forms of business fail because they have not assessed properly and articulate the demand pattern. Many of these businesses are established based on figments of imagination that market exists for goods and services.

A wise entrepreneur will systemically scan the market to see if the level of demand for proposed goods and services would justify his efforts. If the point to the contrary, the decision to launch the enterprises should be dropped.

(b) Economic factor

The dynamic environment in which the business operates is full of threats and opportunities. Efforts should be made to predict the future trend and adjust on anticipation of new development. One should not fold his arms and wait, for it may be late to catch up with changes which the new development brings with it.

(c) Political factor

In Nigeria and other developing countries, the government plays a central role in economic planning. Therefore, a great deal of information may obtain and form periodic and annual development plans, budgets, and other government services. The rate of turnover of political leaders comes with different ideas about how the country should be administers. Where the turnover is high, projects with shorter period are recommended.

(d) Social-Cultural factor

This can be defined as the aggregate of attitudes, beliefs and customs of the people in a given society. An entrepreneur should study the socio-cultural environment way well before reaching the decision to invest. If the people in a given area are predominantly Muslims, establishing a hotel where wine, spirits and other alcoholic beverages is to be sold, will not stand the test of time.

(e) Technological factor

For a business to compete favorably with its competitors, it has to be abreast with latest technology in its area of operation. Technology impact on business in terms of efficiency and effectiveness.

Borrowing a leaf from Large Scale Enterprises, Small and Medium Scale enterpreneurs should embark on strategic planning. Strategic planning areas in a business be it small or big, include the formulation of policies not only on personnel, marketing, production, finance and insurance but also on profitability. Policies emanate from plans. The area general statements are designed to guide the activities of business operations towards the attainment of the corporate goal. It provides the structure on which future activities can be based.

(a) Financial and investment policies

In the course as new business, the financial and investment policies should be given pride of place. They are very important because of their profound effects on profitability and survival of the new business. Considerations should be given to the structure of the firm. Capital structure, sometimes known as financial plan points to the composition of long-term sources of funds, such as debentures, long-term debts, preference shares, capital and equity shares, capital including reserves and surplus.Some businesses do not plan their capital structure; it develops somewhere along the line with the decisions of the financial manager. Such company can start with a shabby arrangement in the short-term, but ultimately, they will encounter difficulties in raising funds to finance their activities.

Therefore, a SME should plan its capital structure to maximize the use of funds. A retained earnings is one significant way of financing a business on the long run; no serious cost is involved in getting this fund. On the other hand, investors deserve at least fair dividend over a period to justify their investments.A balance has to be stricken between a long-term financing decision and wealth maximization decision, Schmoth (1993).

(b) Personnel policies

The central theory in the firm's personnel policy is to find the right man for the right job so that his talent may be polished and the objective of the organization is realized. Due recognition must be given to on-the-job-training. While it may be difficult to train people to spot economic opportunities, it is possible to train them, and assess such opportunities. Similarly, certain managerial skills are trainable, but without them, new SMEs might not survive because of their inability to overcome initial management difficulties.

(c) Marketing policies

Marketing policies, which will give a clear answer to the following questions, should be formulated:

- What are the various channel of distribution?
- What should the pricing and credit policies be?
- What will be the branding and packaging policies?

Sound policies, which can give answers to the questions above, will enhance the success of the business?

1.3.1. Past policies of government on SMEs

The government also set up Vocational Improvement Centre's with the assistance of the Ford Foundation in 1964. The pilot scheme was in Kaduna and the scheme was designed to train artisans through the apprenticeship system for two years. The target population was meant to be the self-employed artisan or the apprentices of small-scale industries. By 1970 it was found that over 80% of the people that benefited from the program were neither self-employed nor from the small-scale industrial sector, according to Chuta (1984).

Also, for the provision of technical, managerial and financial assistance, were the National Open Apprenticeship Scheme (NOAS), the Entrepreneurship Development Programme or the Job Creation Loan Guarantee Scheme, all under the umbrella of the National Directorate of Employment (NDE) which was established in November, 1986.

The National Open Apprenticeship Scheme is meant for young school leavers who have no marketable skills. Placed as apprentices with establishments either in the public or private sector for periods long enough to acquire necessary skill and experience, the government through NDE pays for the training of such participants, at the end of which the trainee goes into self-employment either through loans from family or from government through the NDE.

The Job Creation Loan Scheme is otherwise known as Entrepreneurship Development Programme. The scheme was designed to help fight unemployment among university graduates and graduates of other institutions of higher learning. The scheme aimed at making banking lending – terms more flexible towards young graduates who have neither equity nor collateral.

Furthermore, government organized programmes which were in the form of monetary, fiscal and industrial policy measures at the macro level, and financing arrangements at the micro level. Nnanna (2001) grouped these schemes into three namely:

- Provision of local finance through government agencies – Federal Ministry of Industry, Central Bank of Nigeria (CBN), Nigerian Industrial Development Bank (NIDB), Nigerian Bank for Commerce and Industry (NBCI), Nigeria Export-Import-Import Bank (NEXIM) etc.
- Facilitating and guaranteeing external finance through the World Bank, African Development Bank (ADB), and other multilateral institutions set up to support SMEs.
- The establishment of the National Economic Reconstruction Fund (NERFUND), which was a source of medium and long-term loans to SMEs.

A detailed analysis of the performance of each of the above Programmes/ Schemes is given below:

i. Small-Scale Industries Credit Scheme (SSICS)

This programme was set up in 1971 by the Federal Government through the Federal Ministry of Industry, with the aim of providing financial and technical support for the SMEs. This scheme, which was launched as the Small Scale Industries Credit Scheme (SSICS) in the Third National Development Plan (1975–1980), was operated as a matching grant arrangement between the Federal and State Governments. The scheme was managed by the States' Ministry of industry, Trade and Cooperatives.

Available evidence shows that this scheme was largely unsuccessful due to the dearth of executive capacity to appraise, supervise and monitor projects. This resulted in the funding of many unviable projects with the attendant high loan default rates. Consequently, the scheme, which was expected to be a resolving one, became progressively cash-strapped and moribund in 1979.

ii. Central Bank of Nigeria (CBN) – Directed SME funding

The Central Bank of Nigeria indirectly promoted the financing of SMEs through its credit guidelines, which required commercial and merchant banks to allocate a minimum stipulated credit to the preferred sectors, including the SMEs. For instance, in the 1979/80 fiscal year, the CBN directed that at least 10 per cent of the total loans and advances to indigenous borrowers be allocated to SMEs. This was subsequently raised to 16 and 20 per cent in 1980 and 1990, respectively. However, due to the high default rate coupled with the cumbersome loan administration, banks preferred to pay the prescribed penalties for non-compliance, rather than channel credit to the SMEs.

The imposition of sanctions by the CBN, including mandatory transfer of shortfalls by defaulting banks to the development banks compelled most commercial and merchant banks to comply with the CBN directives, resulting in credit expansion to the SMEs and micro enterprises. Unfortunately, the sectoral credit allocation policy was jettisoned in October 1996, following the full liberalization of the financial sector.

iii. Nigerian Bank for Commerce and Industry (NBCI)

This bank was set up in 1973 to provide financial services to the indigenous business community, particularly the SMEs. The bank approved a total of 797 projects valued at ₦ 965.5 million between 1973 and 1989. However, the bank suffered major operational problems which crippled its operations since 1989.

iv. Nigerian Industrial Development Bank (NIDB)

This bank had a primary mandate of providing credit facilities for the large-sale industries, however, the bank accommodated SMEs with total assets and working capital of up to ₦ 750,000.00. A remarkable aspect of the NIDB financing of SMEs was its policy of equity invepstment in the paid up share capital of some of the projects it financed. It is on record that the bank disbursed a total of ₦ 174.6 million to SMEs between 1980 and 1988. It was also responsible for disbursing most of the funds under the SME II loan scheme. However, due to a number of constraints, the bank's involvement in the financing of SMEs dwindled in the late 90s. as a result of the operational problems faced by NIDB and other development banks, it has now been merged with NBCI and NERFUND to form the Bank of Industry.

v. World Bank SME II loan

The Federal Government secured a World Bank loan of $270.0 million in 1988, aimed at improving the quality and range of financial and extension services available to SMEs. The Scheme's loans, which were for start-up, expansion and rehabilitation of existing enterprises, were mainly targeted at enterprises in manufacturing, printing, mining, agriculture, agro-allied, quarrying, transport, and leasing. An apex SME Unit was established in the CBN in 1990 to administer the credit component for the World Bank in order to facilitate project implementation. Available records show that the SME Apex Office approved a total of 211 projects valued at US$132.8 million between 1990 and 1994 when the scheme was abolished, out of which US$106.5 or 80 per cent was disbursed.

vi. National Economic Reconstruction Fund (NERFUND)

This scheme was introduced in 1989 during the Structural Adjustment Programme to assist SMEs cope with the attendant high production costs arising from high cost of imported inputs and high rates of interest. In addition, NERFUND was established to bridge the long-term financing gap in banks' lending to SMEs. The fund was expected to provide long-term funds for wholly Nigerian owned SMEs operating in the area of manufacturing, agro-allied enterprises, mining, quarrying, industrial support services, equipment leasing, and other ancillary projects. The resources of NERFUND were mainly contributions from the Federal Government, the CBN, and foreign sources. Available records show that from inception in 1989 to 1998, NERFUND disbursed about US$144.9 million and ₦ 681.5 million to the approved projects. In spite of the success recorded by the fund, it faced a number of problems, such as poor loan recovery, dwindling demand for loans due to foreign exchange risks, under funding and uncooperative attitude on the part of some participating banks. On the part of the SMEs operators, the collateral requirements stipulated by the participating banks were considered to be very stringent, as they limited SMEs' ability to access the loan facilities.

vii. The capital market

The Second Tier Securities Market (SSM) was established in 1985 to assist SMEs access funds from the capital market mainly for the purpose of business expansion and modernization. This initiative was necessary to counter the bias of the capital market in favour of large-scale industries. Available records show that the number of SMEs listed on the SSM rose to about twenty, while 4 of such firms were upgraded to the First Tier Market.

Table 3: Ratio of loans to small scale enterprises to commercial banks' total credit (1992-2014)

Year	Commercial Banks Loans to Small Scale Enterprises (N'm)	Commercial Banks Total Credit (N'm)	Commercial Banks Loans to Small Scale Enterprises at percentage of Total Credit
1992	20,400.0	41,810.0	48.8
1993	15,469.9	48,056.0	32.2
1994	20552.5	92,624.0	22.2
1995	32,374.5	141,146.0	22.9
1996	42,302.1	169,242.0	25.0
1997	40,844.3	240,782.0	17.0
1998	42,260.7	272,895.5	15.5
1999	46,824.0	353,081.1	15.3
2000	44,542.3	508,302.2	8.7
2001	52,428.4	796,164.8	6.6
2002	82,368.4	954,628.8	8.6
2003	90,176.5	1,210,033.1	7.5
2004	54,981.2	1,519,242.7	3.6
2005	50,672.6	1,899,346.4	2.7
2006	25713.7	2,524,297.9	1.02
2007	41,100.4	4,813,488.8	0.85
2008	13,512.2	7,806,751.4	0.17
2009	63,300.7	35,167,203.7	0.17
2010	60,424.7	37,433,800.6	0.16
2011	62,145.8	38,546.450.4	0.18
2012	65,661.2	38,678,302.1	0.24
2013	68,578.6	40,089,222.4	0.53
2014	72,567.5	40,679,567.5	0.59

Source: CBN Statistical Bulletin – Computed from Deposit Money Bank's Returns

The amount of percentage of total loan to small scale enterprises as a proportion of total credit reduced drastically over the years. As a matter of fact the total loan to small scale enterprises reduced drastically by over 50% in 2006 with total credit increased by nearly 50%; talk less of 2010. This shows that SMEs is no more a priority.

As such, it will be unusual to judge Nigeria's economic performance by comparing it with macroeconomic outcomes with those of the developed economies in this research work. That is not what we intends to do; what we intend to do is to look at the macroeconomic performance in Nigeria relative to some selected African countries using the GDP per capita at purchasing power parity, real GDP growth, poverty and the Human Development Index (HDI).

Chapter 2

Medium and Small Scale Enterprises to Socio-Economic Development

2.1. Place of SME in Remeding The Nigerian Economic Paradox

This can best be explained by the *two-gap economic model*. The Two-Gap Models of development are contained in the Post-Keynesian growth models designed by Harrod and Domar. They tried to identify the pre-conditions for the economic growth of market economies. These two preconditions are essentially rooted in the Nigerian economy and these are (1) Internally: inadequate savings would definitely translate to investment. The GAP between these two is called saving constraints (SAVING GAP). Closing this gap requires SME-financing. (2) Externally: inadequate foreign exchange arising from inability to export vis-à-vis high importation will lead to short fall in foreign exchange. The GAP between these two is called foreign exchange constraints (TRADE GAP) which can be corrected by facilitating the establishment of indigenous enterprises through a national embrace of the SME-initiatives. Nigerian leaders resorted to foreign aid assistance from developed countries, multilateral and bilateral international organizations to salvage the economy from total collapse; but, we can't continue to carry the begging bow as our benefactors only give us fish and don't teach us how to catch fish.

Therefore, to emphasize the need for economic growth and development, there is need to diversify the economy from oil exploration and exploitation to productive effort of agricultural and other real sectors through an SME-initiative. However, inability to revive these sectors spells doom to the trade and investment platform of Nigeria.

As such, to do justice to the impact SMEs can make in alleviating unemployment, *Robert Owen's philosophy* (of job creation for the masses and improvement of the standard of living of the working class through the creation of villages of cooperation) and *Alfred Marshal's philosophy* (which states that labour is the true source of all value) will be synergized. A mixture of two schools of thoughts: socialism and neo-capitalism. Then, the ideas of these synergized philosophies will be applied to the Nigerian situation. Also, a critical examination of the concept of *mercantilism*; thereby inspiring initiatives on roles key stakeholders (i.e. governments, large scale enterprises and private individuals) should play in promoting SMEs and alleviating unemployment.

Table 1: Some benefits and costs of Export, Import, Provision of SME Loans, and Migration

	Benefits	Costs
Export	• Increased income • Wider supply base	• External shocks
Import	• Access to intermediate materials	• External shocks
SME Loans	• Domestic investment augmentation • Skill Acquisition • Productivity growth	• Financial market crisis
Migration	• Brain gain • Home remittances	• Brain drain

Source: Generated by the Researcher

Taking a close interest in migration on table above, the impact of migration on an economy depends on the potential contribution of those emigrating in terms of their marginal physical product and that of their home remittances. These impacts, however, need to be juxtaposed with the loss of well-educated and highly productive labour force. This is the mystery behind why Nigeria which was far ahead of Singapore in Macroeconomic outcomes in the '60s is far behind. While we suffered brain drain over the years, Singapore enjoyed brain gain after putting in place all that was necessary for economic productivity.

2.2. The Contributions of Small and Medium Scale Enterprises to Socio-Economic Development

It is incontrovertible that small and medium-scale industries are the bedrock of development in any nation state. This is sequel to the fact that many large businesses grew out of small and medium enterprises. The experience of the industrialized nations suggests that the development of SMEs is a precursor to the development of the industrial sector. For instance, available evidence shows that the industrial nations of North America and Western Europe started with the development of their small and medium-scale industries. The prominence given to the SMEs in those countries is demonstrated by the decisive steps taken to stimulate and sustain the growth of SMEs in those countries. For instance, the United States of America enacted The Small Business Act in 1953. In addition the President of the United States of America is required by law to give a State of Small Business Address similar to the State of the Union Address annually. Similarly, the Netherlands issued its white paper on the development of small-scale industries in 1954. The Federal Republic of Germany followed suit with its Policy Framework on SMEs, which was enacted in1959, while Japan in 1963 enacted the Fundamental Laws on SMEs. It is therefore not surprising that the SMEs contributed immensely to the success story of the industrialization process in these countries.

In an analysis of the contributions of SMEs to the socio-economic development of an economy, Owualah (1999), identified the following functions of SMEs:

- Transformation of traditional or indigenous industry;
- Stimulation of indigenous entrepreneurship and technology;
- Creation of jobs;
- Redistribution of wealth and income;
- Utilization of local resources;
- Dispersal and diversification of economic activities; and
- Mobilization of savings.

i. Transformation of traditional/indigenous industry

Evidence from both the developed and developing countries show that the traditional sector always precedes the modern sector. To a large extent, the modern sector evolved through a structural transformation and modernization of the traditional sector. The critical driver of this change has always been the small firms, such as household and artisan industries which metamorphose into medium and large-scale firms as the operators acquire new skills and their scale of operation expands.

ii. Stimulation of indigenous entrepreneurship

Through the seedbed or nursery role, SMEs provide opportunities for the realization of the innate entrepreneurial endowments aside from servicing as vehicles for the propagation and diffusion of innovative ideas. Furthermore, SMEs provide opportunities for imported technologies to be tried before they are adopted to suit the local condition.

iii. Employment creation

A very important role of SMEs their potential to create jobs due to their labour-intensive and capital-saving mode of operations. A study by Birch (1979) on the manufacturing and private sector service establishments in the United State of America shows that two-thirds of new jobs were created in firms employing less than twenty workers. Sixty-six per cent of these new jobs were crated by new, young, small independent firms. In the case of the developing countries, the labour-intensive nature of SMEs coupled with the death of capital and abundance of labour presents the SMEs as a veritable source of employment.

iv. Wealth redistribution

By providing employment and remunerative economic activities to a great number of rural and urban people. And also supplementing their income from their regular jobs, SMEs contribute greatly to the reduction of income disparities.

v. Utilization of local resources

SMEs use a lot of local raw materials and discard by-products of large firms as-primary inputs in their production process. In addition, their employment of local resources could be seen from the perspective of the opportunities being offered by SMEs to people with limited skills and formal training to acquire new skills or improve on their old skills.

vi. Dispersal of economic activities

Since SMEs are not capital-intensive, they spring up anywhere in response to the demand for their products. Furthermore, they have the tendency of being located outside major metropolitan centers thereby contributing to industrial dispersal.

vii. Mobilization of savings

SMEs are useful in mobilizing small savings for productive investments and enhancement of the industrial capital formation. By channeling personal savings of the promoter and those of his or her friends into the production process, SMEs are able to transform the savings into the production of physical goods and services thereby bring about the expansion of the funds due to the multiplier effect.

2.3. Basic Record Keeping and Accounting for Small Business

Record keeping is vital for effective financial management for small businesses. All records of transactions are expected to be thoroughly kept. The major record books (source documents) kept by small businesses include:

(a) ***Sales Day Book*** – This is used for recording sales transactions that are not promptly received. That is sales made on credit. Sales paid for will be recorded in the cash book.

(b) ***Purchase Day Book*** – This is used to record goods bought on credit. Purchases paid for will also be recorded in the cash book.

(c) ***Return Outward Book*** – This is used for recording goods bought but later returned due to damages, wrong brands, colour, etc.

(d) ***Return Inward Journal*** – This is used for recording goods sold but later returned due to wrong supply of products, damages, etc.

(e) ***Debit Note*** – This is used to record differences where the customer is undercharged. The note will be sent to the buyer to show that he has been undercharged and that his account has been debited with the amount.

(f) ***Credit Note*** – This is issued where a client has been overcharged. This is sent to the customer to show that his account has been credited with the amount of overcharging.

(g) ***Receipt*** – This is used for acknowledging money received. It serves as evidence of payment made.

(h) ***Journal Proper*** – This is used to record other transactions which could not be conveniently recorded in other source document. This includes purchases of fixed assets on credit, sales of fixed asset on credit, opening balance and closing balance of an account. The information recorded in these source documents can then be used in preparing the trading profit and loss account as well as the balance sheet. A conducive and enabling environment is essential for the survival of new small businesses.

The overall macro and micro economic framework, the legal and regulatory environment as well as political climate influence the rate of creation of new businesses, their survival and growth. Small business will develop when there is a supportive environment.Small businesses because of their size and characteristics have certain constraints that require them to seek outside support. To offset these constraints, and disadvantages, government often give them some support programmes. These programmes and institutions provide on need basis services that are meant to assist the small business in the areas of development of internal management system, improving productivity, technology upgrading, etc. The new entrepreneurs need to seek further information on these support programmes. It will assist them in the planning of their businesses.Indeed, lending to small and medium-sized enterprises (SMEs) in particular has not taken off in advanced economies. In the euro area in particular, lending conditions have remained tighter than before the crisis despite a return importance of SMEs in generating investment and employment, going forward it will be crucial to relieve their financing conditions and allow them more broad-based access to banking and market- based credit. In part, such an improvement in financing conditions can be achieved by speeding up the implementation of the announced and agreed banking sector reforms to help to transform the current banking sector model and make it more amenable for real economy financing.

In this regard, it should be stressed that proper and comprehensive financial sector regulation can actually contribute to faster employment growth It will relieve enter- prizes and banks from economic and regulatory uncertainty and put the business model of the banking sector on a more stable footing. The reduced volatility in domestic and international markets that such tighter regulation might induce is a prime requisite factor for stimulating both investment and employment growth and might help to reduce precautionary saving. In addition, stricter prudential regulation and the limitation of implicit public guarantees against bank failures will help phase out current exceptional monetary measures, restoring market forces in the banking sector. This will improve financial conditions in the real economy, as banks will have greater incentives to channel their funds toward productive ends rather than volatile financial products. Adding up these effects, estimates by the ILO show that broad- based financial sector regulation could add more than half a percentage point to job creation rates (ILO, 2011a).

2.4. Challenges of Entrepreneurship in a Growing Economy

Nigeria as a growing economy is not exempted from the challenges peculiar with growing economy and nations. Entrepreneurs or small-scale businesses are normally faced with daunting but surmountable challenges which at times are two much for many of them to succeed.

Entrepreneurial ventures are saddled with lots of challenges to include:

(i) Insecurity

Nigeria has a porous security system, where lives and property are not safe and thereby leaving the citizens to live in perpetual fear. In recent times this insecurity took a different turn, with the advent of kidnap of foreigners now followed by indigenes with the bid to get ransom paid to them (kidnappers). Theft, stealing and broad day light robbery is on the increase.

(ii) Lack of stable power supply

It is sad to note in the 21st century, for a country as blessed as Nigeria to still grope in darkness which has become a way of life. The epileptic nature of power makes business owners resort to alternative uses of power thereby contributing to the cost of transacting business in Nigeria.

(iii) Incessant change in government policies

Political problems: No stability in the government, constant change of government.

(iv) Technological problems

Any nation that is not technologically driven lives at the mercy of other nations. Developed nations dictate what and how to produce.

(v) Infrastructural decay

Bad roads and other amenities to enhance business activities have all collapsed.

(vi) High tariff/inflation

A good example is the advent of MTN to Nigeria. The then Nigerian Government heavily tasked the communication service providers (MTN) as operational permit who in turn charged the customers accordingly.

(vii) Economic problems

Most business owners hardly can access loan facilities from banks and are therefore left to raising money either by personal savings or their nations.

Chapter 3

Promotional Entrepreneurship Programmes in Nigeria

3.1. Approaches for Promotion Entrepreneurship Programmes in Nigeria

This book used the following typology of approaches:

1. Start: Stimulating local business start-ups that enter into new local markets, there - by avoiding the trap of yet another 'me-too' enterprise with low revenues. Promote business start-ups as a way of creating additional jobs and incomes, and promoting competition and structural change. However, these startups often lack innovative business ideas. Instead they enter sectors with a low entry threshold, but with high competition from other companies. Stimulating local start-ups that apply IB models could encourage players to enter new markets that are not yet saturated with competitors, while at the same time serving development goals. Possible target groups for this approach include university students, NGOs or experienced business people who create a spin-off company or leave their company to engage in a totally new business. In a country or sector where IB models are still rare, the first questions that arise are why this is the case, and whether potential markets exist for the use of IB models. Questions to ask include

- What products and services do poor people need, and which of these would they be willing to pay for, given the appropriate assistance? What are the quality and quantity (outreach) of public services being delivered to society? Where is there room for improvement, and for private sector initiatives?
- Is the government driven by a desire for social impacts? Which sectors, regions or groups do government policies focus on?

- How the sectors are regulated – especially water, energy, health, education, financial services or agriculture? Does the government allow private sector activity and competition in these markets; what is its general attitude towards the private sector?
- Are companies aware of the potential for integrating the BoP in their business models? What is the culture of entrepreneurship, and where do possible 'hubs' for entrepreneurship exist?
- Are there any active NGOs, civil society groups or other (profit-based) intermediaries that could act as partners?
- What are the social, environmental and economic trends in the country?

2. Adopt: Encouraging local companies to adopt and encourage integrating poor and small-scale producers, suppliers, distributors and workers into their value chain, thereby contributing to employment and income creation, or by designing new products and services for poor customers. Such programmes can be conducted in a specific sector or more generally. They can be initiated and funded by donors together with local organisations (business organizations, universities, etc.) at a relatively modest cost. For example, the SEED Initiative of UNEP, UNDP and IUCN runs award competitions for innovative start-up ventures that address social and environmental challenges in Africa, Asia, Eastern Europe and Latin America. In an annual competition, locally based social and environmental entrepreneurs may apply. In addition to prize money, award winners benefit from business support services (e.g. Business plan support), access to relevant networks and high-level profiling.

3. Scale: Scaling up existing models to achieve broader outreach and a greater development impact, and to foster the growth of these companies, thereby creating employment and affecting incomes. The approach used for promoting the domestic market will largely depend on:

- The development priorities of the partner with regard to certain sectors (e.g. health, education, energy), regions (disadvantaged regions or rural areas) or groups (such as unemployed youth, disadvantaged or formerly disadvantaged groups, ex-combatants *etc.*)
- The level and direction of demand and market development for certain products and services
- The structure of the private sector and the environment for entrepreneurship programmes in the respective country (e.g. existence of large local companies)
- The programme's overall goals and strategy
- The collaboration opportunities with other stake - holders and donors.

3.2. Appraisal of Past Policies of Government at Promoting Entrepreneurship

Beginning with the Industrial Development Centres (IDCs) which were started in 1964, and located at Zaria, Owerri and Osogbo with the assistance of the government of the United States of America, the IDCs were supposed to provide technical, managerial and accounting assistance to SMEs in especially five major areas – wood working, metal working, automobile repair, textiles and leather-working. The IDCs consisted of workshops for demonstration and training, where entrepreneurs were to be familiarized with new production techniques, new types of machinery, etc.

3.3. Strengths and Weaknesses of Small and Medium Enterprises

In addition to the general barriers to doing business with the poor, organization-specific strengths and challenges need to be considered. These can be categorized as 'local' or 'foreign', 'small' or 'large'. All types of company have their different strengths and weaknesses with regard to applying IB models. Due to their strengths and weaknesses, companies – local or foreign, small or big – all differ in their ability to apply the various strategies to overcome the barriers described above:

- Local SMEs have an advantage when it comes to understanding the needs of low-income clients, and they also benefit from being well anchored in the local community. Their core business often involves selling to, and sourcing from low-income populations. This means they are adept at developing low-cost products and services as well as adequate distribution strategies.
- On the other hand, SMEs often have little access to adequate human, information and financial resources, and lack strategic planning capacities. Individuals and SMEs with innovative ideas rarely possess the right skills to develop a sophisticated business plan, and it may be hard to attract financing in the early stages due to the high risk of new, unproven IB ventures. Related problems occur when the amount of capital needed falls below the investor's threshold, or when there company lacks collateral. At present, therefore, from the point of view of investors there seems to be a shortage of feasible, replicable IB opportunities. This is corroborated by the fact that in some countries there is an abundance of capital ready to finance IB ventures, although there is a dearth of applications based on proven, sufficiently large and financially sound business models.

- Moreover, having little access to technologies and information beyond the local context can hinder a company's capacity for innovation. Operating with a low profit margin can slow down growth processes and make it difficult to scale up operations.
- Large national companies or emerging market multinationals seem quite well positioned to invest in IB models. They combine a good general understanding of clients' needs, established local supplier networks, widely known brands, and adequate human and financial resources. They are used to conducting business in a difficult environment, and national companies in particular are well-connected locally.
- However, such companies might still support in increasing the social value and the 'inclusiveness' of their products and services, in finding civil society partners, in further adapting their products and services to customers' needs, and in unlocking additional financing or engaging in the policy dialogue. As the potential partners and suppliers for doing business with the BoP tend to vary from country to country, large national companies or emerging market multinationals will need support to establish strong networks when entering new markets.
- Multinational companies from industrialized countries are at the other end of the spectrum. They tend to be well equipped with human and financial resources, as well as R&D capacities. This means they are able to make more substantial investments to overcome barriers, adapt products and services, and engage in policy dialogue.
- However, they are often unfamiliar with the ideas and approaches of IB models, and they might question the effectiveness of such models for them, particularly when they are pursuing short-term objectives to maximize shareholder value. The concept of inclusive business often runs counter to established internal processes and standards, such as minimum profit margins or maximum investment horizons.
- Other challenges include the lack of market information, and poor understanding of the peculiarities of the low-income target market (such as consumption patterns, informal competition, and weak supply structures). These larger multinationals also encounter practical challenges in implementation, such as language barriers or cultural differences. Private sector development (PSD) programmes wishing to focus on the wider use of IB models in order to achieve (private sector) development objectives can work with any of these types of company. They all have the potential to improve living conditions for the poor and to increase productive capacities, incomes and employment in developing countries.

Rooted in local community and networks Good understanding of local culture, clients' needs and business environment Adopt at developing low-cost goods and services. These are:

(a) Small

High degree of flexibility and potential for radical innovation local networks built on trust.

(b) Foreign

Broad knowledge base Good access to international support networks.

(c) Large

Strong financial and human resource base policy-level influence strong brand.

Chapter 4

Issues in Employment in Nigeria

4.1. Issues in Employment

4.1.1. What is employment?

Employment is a relationship between two parties, usually based on a contract, one being the employer and the other being the employee.

(a) Employee

An employee contributes labor and expertise to an endeavor of an employer and is usually hired to perform specific duties which are packaged into a job. In most modern economies, the term "employee" refers to a specific defined relationship between an individual and a corporation, which differs from those of customer or client. Other types of employment are arrangements such as indenturing which is now highly unusual in developed nations but still happens elsewhere.

(b) Employer-worker relationship

An employer's level of power over its workers is dependent upon numerous factors, the most influential being the nature of the contractual relationship between the two. This relationship is affected by three significant factors: interests, control and motivation. It is generally considered the employers' responsibility to manage and balance these factors in a way that enables a harmonious and productive working relationship. Employer and managerial control within an organization rests at many levels and has important implications for staff and productivity alike, with control forming the fundamental link between desired outcomes and actual processes. Employers must balance interests such as decreasing wage constraints with a maximization of labour productivity in order to achieve a profitable and productive employment relationship.

(c) Finding employees or employment

The main ways for employers to find workers and for people to find employers are via jobs listings in newspapers (via classified advertising) and online, also called job boards. Employers and job seekers also often find each other via professional recruitment consultants which receive a commission from the employer to find, screen and select suitable candidates. However, a study has shown that such consultants may not be reliable when they fail to use established principles in selecting employees. A more traditional approach is with a "Help Wanted" sign in the establishment (usually hung on a window or door or placed on a store counter).

(d) Workforce organizing

Employees can organize into trade or labor unions, which represent the work force to collectively bargain with the management of organisations about working and contractual conditions.

(e) Ending employment

Usually, either an employee or employer may end the relationship at any time. This is referred to as at-will employment. The contract between the two parties specifies the responsibilities of each when ending the relationship and may include requirements such as notice periods, severance pay, and security measures.

(f) Employment contract

(1) Australia

Australian Employment has been governed by the Fair Work Act since 2009.

(2) Bangladesh

Bangladesh Association of International Recruiting Agencies (BAIRA) is an association of national level with its international reputation of co-operation and welfare of the migrant workforce as well as its approximately 700 member agencies in collaboration with and support from the Government of Bangladesh. The BAIRA Exporting skill manpower in different countries of the world.

(3) Canada

In the Canadian province of Ontario, formal complaints can be brought to the Ministry of Labour (Ontario). In the province of Quebec, grievances can be filed with the Commission des normes du travail.

(4) Pakistan

Pakistan has Contract Labour, Minimum Wage and Provident Funds Acts. Contract labour in Pakistan must be paid minimum wage and certain facilities are to be provided to labour. However, a lot of work has yet to be done to fully implement the Acts.

(5) India

India has Contract Labour, Minimum Wage, Provident Funds Act and various other acts to comply with. Contract labour in India must be paid minimum wage and certain facilities are to be provided to labour. However, a lot of work has yet to be done to fully implement the Act.

(6) Philippines

In the Philippines, private employment is regulated under the Labor Code of the Philippines by the Department of Labor and Employment.

(7) United States

In the United States, the standard employment relationship is considered to be at-will, meaning that the employer and employee are both free to terminate the employment at any time and for any cause, or for no cause at all. However, if a termination of employment by the employer is deemed unjust by the employee, there can be legal recourse to challenge such a termination. Unjust termination may include termination due to discrimination because of an individual's race, national origin, sex or gender, pregnancy, age, physical or mental disability, religion, or military status. Additional protections apply in some states, for instance in California unjust termination reasons include marital status, ancestry, sexual orientation or medical condition. Despite whatever agreement an employer makes with an employee for the employee wages, an employee is entitled to certain minimum wages set by the federal government. The states may set their own minimum wage that is higher than the federal government's to ensure a higher standard of living or living wage for their residents. Under the Equal Pay Act of 1963 an employer may not give different wages based on sex alone.

Employees are often contrasted with independent contractors, especially when there is dispute as to the worker's entitlement to have matching taxes paid, workers compensation, and unemployment insurance benefits. However, in September 2009, the court case of *Brown v. J. Kaz, Inc.* ruled that independent contractors are regarded as employees for the purpose of discrimination laws if they work for the employer on a regular basis, and said employer directs the time, place, and manner of employment.

According to Swedish law, there are three types of employment.

- Test employment (*swe: Provanställning*), where the employer hires a person for a test period of max 6 months. The employment can be ended at any time without giving any reason. This type of employment can be offered only once per employer and employee. Usually a time limited or normal employment is offered after a test employment.
- Time limited employment (*swe: Tidsbegränsad anställning*). The employer hires a person for a specified time. Usually they are extended for a new period. Total maximum two years per employee per employer and employee, then it automatically counts as a normal employment.
- Normal employment (*swe: Tillsvidareanställning / Fast anställning*), which has no time limit (except for retirement etc.). It can still be ended for two reasons: personal reason, only strong reasons such as crime. Or: lack of work tasks (*swe: Arbetsbrist*), cancellation of employment, usually because of bad income for the company. There is a cancellation period of 1–6 months, and rules for how to select employees, basically those with shortest employment time shall be cancelled first.

There are no laws about minimum salary in Sweden. Instead there are agreements between employer organisations and trade unions about minimum salaries, and other employment conditions.

4.2. Unemployment in Nigeria

Unemployment, by default, is the difference between the lab our gainfully employed at the wages and working conditions that exist, and the amount of labour available, however, Gbosi (2006) defined unemployment as a situation in which people who are willing to work at the prevailing wage rate are unable to find jobs. This implies that not just anyone who is to be counted as part of the unemployed labour force, in order to a void overestimation of the official rate of unemployment. The International Labour Organization (ILO) defines unemployed in this manner, "the unemployed is a member of the economically active population, who are without work but available for and seeking for work, including people who have lost their jobs and those who have voluntarily left work (World Bank, 1998). The application of this definition across countries has been faulted, especially for the purpose of comparison and policy formulation, as countries characteristics are not the same in their commitment to resolving unemployment problems (Douglason *et al.*, 2006).

The rate of unemployment in Nigeria can be attributed to a lot of factors, including the depression in the 1980s and during the late 1970s. Economic downturn leads to the implementation of stabilization measures which includes restriction on exports, to increase dependency on Nigerian manufacturing enterprises and the resultant effects are mostly not positive. The analysis by educational status in past years suggests that people who have been majorly affected by unemployment are those without basic education, however today, even the educated have acute troubles getting employed. It is impressive to note here that, in 2003, Nigerian's unemployment rate declined substantially to 2.3 percent.

This decline was attributed to the various government efforts aimed at addressing the problem through poverty alleviation programmes. This decline also pointed to an increased number of people who got engaged in the informal sector activities. They also found that education of owner of a business enterprise was a significant factor influencing efficiency. They conclude that the evidence of variations in efficiency is indicative of the need for more proactive actions to raise the level of efficiency and employment among the firms in the sample. Employment generation has been seen as a means of alleviating poverty, increasing the level of economic activities which translate into economic growth. The situation of unemployment in Africa, Nigeria as a case study has been on the increase which has resulted in increase in social vices among other negativities. Although the Nigerian Government put in place policies and programmes which are meant to combat this menace, but the inadequacies of implementation and up till now these programmes have not made much impact. According to Bloom (2000); if Nigeria is able to overcome its challenges and collect its demographic dividend, we estimate that:

- Nigeria can obtain almost 12% increased GDP per capita over the default scenario by 2020, and more than 29% increased GDP per capita by 2030. With additional modest institutional improvements, the extra GDP per capita over the default scenario could be almost 13% by 2020 and about 31% by 2030.

- With the demographic dividend, Nigeria' s economy can be 3 times larger than today in 2030, instead of only 2 times larger without the demographic dividend.

- By realizing the demographic dividend, Nigeria can lift about 5.8 million more people out of poverty by 2020 and about 31.8 million by 2030, over and above the default scenario. With institutional improvements, the number of additional people lifted out of poverty can be 34 million by 2030.

- By increasing investment in human capital as fertility rates decline, Nigeria could increase GDP by nearly 50% and sustain that gain indefinitely. But to realize its demographic dividend, we estimate that Niger ia will need to create around 24 million new jobs in the next decade and around 50 million new jobs over 2010-2030. Further, the jobs will have to be productive. This will require increasing Nigeria's human capital which cannot be achieved without strong investments in health, education, gender parity, and institutions. In particular, the role of education is so closely tied together with health, gender parity and institutions that any specific intervention that ignores the other three aspects is very unlikely to work. Even if some benefits are realizable in the short term, they will not be sustainable.

According to the Northern Illinois University Outreach (NIU, 2005), Higher education has historically included economic development as part of its core mission. Colleges and universities have allocated fiscal, physical, and human resources and created entrepreneurship systems within the institutions to advance economic development. Senior administrators provide strong, visible leadership designed to:

- Create a quality workforce by growing, training, and attracting the finest talent.
- Support current business and industry.
- Improve learning and teaching from pre-school through graduate school.
- Take strong and visible roles in regional initiatives.
- Disseminate research and promote technology transfer.
- Enhance the technology infrastructure.
- Promote livable communities.
- Employ a diverse workforce.

4.3. Vulnerable Groups in Africa

(a) Poor

- Although aggregate data remain subject to wide margins of error, the World Hunger Report estimates that the world's poor fall into roughly three categories. About 60% are small farmers and pastoralists, 20% are landless rural laborers and nonfarm entrepreneurs, and the remaining 20% urban poor (Sanchez and Swaminathan, 2005).
- The rural poor include a heterogeneous mix of economic livelihoods, genders and ethnicities (Von Maltzahn and Durrheim 2008). Small farmers and pastoralists account for the majority of rural poor in most settings, particularly in rural Africa, although even their nonfarm earnings

account for roughly 25% of rural incomes (Haggblade, Hazell and Reardon, 2007; IFAD 2011).

- The principal resource of the rural poor is unskilled labor. Given generally low levels of education, they attain consistently lower literacy and numeracy rates than the rural non poor. In general, they have larger families as well as less land, livestock and access to clean water than the non poor. Low levels of human, physical, financial and social capital operate as frequent constraints to their economic mobility and condition their opportunities in specific value chains.

(b) Women

- Women, like the rural poor, are a heterogeneous group. Age, marital status, religion, ethnicity, family size, location, culture and politics all influence their economic and social status (Quisumbng and Maluccio 2000). Married women and many widows may benefit from larger support networks and asset bases than do young, unmarried and abandoned mothers. But in polygamous families, big differences exist between first wives, second wives and third wives. So even the general category of "married" women includes a highly heterogeneous grouping.
- Many women, nonetheless, face special constraints on their mobility and on their time. They typically assume responsibility for child care, for feeding the family and for ensuring water and fuel supplies. Hence, they are time-poor and often with limited access to land. They typically work in their own fields, in their husband's fields and many times in home-based, labor-intensive nonfarm pursuits. They often face much more difficult access to credit than men, and in some cases have more difficulty than men in establishing title to land. Compared to men, rural women are less likely to be organized in marketing associations.

(c) Youth

- In the short run, high fertility rates among poor households translate into larger families, higher dependency ratios. In the medium term, high fertility gives rise to a demographic bulge of youth (between the ages of 15 and 24) whose numbers pressure labor markets, land access and social systems. The challenge of this emerging youth bulge remains particularly acute in Africa.
- Given continued high fertility rates, the African continent remains the youngest in the world. As a result, youth (between the ages of 15 and 24) represent 40% of the workforce and 60% of the unemployed (World Bank 2009). In rural areas, youth account for about two-thirds of agricultural employment (ILO 2007).

- The growing demographic bulge suggests that youth employment will constitute one of the continent's key challenges over the coming decades. Young job seekers face generation-specific challenges, particularly difficulties in accessing land, livestock and other productive assets while their parents continue to farm actively. Under Africa's ruling rural gerontocracy, those holding power may not be in formal leadership positions. Growing demographic pressure on farmland and pastures restrict young adults' futures in agriculture. In general, these pressures make high-value, land-conserving agricultural products such as horticulture and poultry increasingly attractive.
- Nonetheless, young workers do enjoy some advantages over older workers. They are typically more mobile, physically stronger and better able to adapt to new information and communication technologies. Moreover, they enjoy longer productive working spans over which to amortize training and education investments. In many agricultural value chains, youth occupy specific entry- level niches, as apprentices, assembly traders and porters, from which they can accumulate savings and transition over time.

4.4. Entrepreneurial Education and its Consequence in Nigerian Economy

Entrepreneurship is not a new concept in Nigeria, however, as much as it is popularly discussed, it can be figuratively stated that it is being poor ly or rarely implemented. As it is, according to Omolayo (2006) is the act of starting a company, arranging business deals and taking risks in order to make a profit through the education skills acquired. Another view of entrepreneurship education is the term given to someone who has innovative ideas and transforms them to profitable activities. Summarily, entrepreneurship can be described as "the process of bringing together creative and innovative ideas and coupling these with management and organizational skills in order to combine people, money and resources to meet an identified need and create wealth. In the same vein, Nwangwu (2007) opined that entrepreneurship is a process of bringing together the factors of production, which include land, labour and capital so as to provide a product or service for public consumption. However, the operational definition of entrepreneurship is the willingness and ability of a person or persons to acquire educational skills to explore and exploit investment opportunities, establish and manage a successful business enterprise.

4.4.1. Who is an Entrepreneur?

Entrepreneur can be defined as an innovating individual who has developed an ongoing business activity where none existed before. Meredith (1983) defined an entrepreneur as a person or persons who possess the ability to recognize and evaluate business opportunities, assemble the necessary resources to take advantage of them and take appropriate action to ensure success. Entrepreneurs are people who constantly discover new markets and try to figure out how to supply those markets efficiently and make a profit. He is a person that searches for change, responds to change, and exploits change by converting change into business opportunity.

4.4.2. Objectives of Entrepreneurial Education

Entrepreneurial education according to Paul (2005) is structured to achieve the following objectives.

- To offer functional education for the youth that will enable them to be self-employed and self-reliant.
- Provide the youth graduates with adequate training that will enable them to be creative and innovative in identifying novel business opportunities.
- To serve as a catalyst for economic growth and development.
- Offer tertiary institution graduates with adequate training in risk management, to make certain bearing feasible.
- To reduce high rule of poverty.
- Create employment generation.
- Reduction in rural-urban migration.
- Provide the young graduates with enough training and support that will enable them to establish a career in small and medium sized businesses.
- To inculcate the spirit of perseverance in the youths and adults which will enable them to persist in any business venture they embark on.
- Create smooth transition from traditional to a modern in dustrial economy.

4.5. Entrepreneurship in Higher Education is Critical to Development

It is impossible to consider economic development or reduction of unemployment in Nigeria, without first considering the indices that create its existence. Primarily; the curriculum, the system of teaching, and execution of practical learning gives an introductory idea of the fact that the Nigerian graduate may be prepared to serve, but not to create. Many courses in Nigerian Universities have nothing to do with business management, development or business start up; courses taken in Sciences and Arts predominate on the technicalities of the courses and not the direct application for profitability outside the walls of the institution. Competitiveness, innovation and economic growth depend on being able to produce future leaders with the skills, attitudes and behaviour to be entrepreneurial and to act at the same time in a socially responsible way. Entrepreneurship is not only about creating business plans and starting new ventures. It is also about creativity, inn ovation and growth, a way of thinking and acting relevant to all parts of the economy and society as well as the whole surrounding ecosystem. This interdependence comprises both institutional rules and environmental conditions that define the range of socially and economically viable entrepreneurial opportunities and the way in which entrepreneurs and other stakeholders shape these surrounding institutional and environmental conditions. It is important to start as early as possible at all levels of formal and informal education. It should be integrated into the education system of primary and secondary schools as well as higher and further education. For effective entrepreneurship education, the curricula over the years must be consistent and coordinated and entrepreneur ship education should continue at higher education institutions. Entrepreneurship programmes and modules can have various objectives, such as:

- Developing entrepreneurial drive among students (raising awareness and motivation).
- Developing the entrepreneurial ability to identify and exploit opportunities.
- Training students in the skills they need to set up a business and manage its growth (European Commission, 2008). Without the influence of education in introducing entrepreneurship to the average Nigerian student, it will be difficult for unemployment to be reduced in the country, as it is important to encourage students to think and act entrepreneurially as well as ethically and socially responsible. The importance of higher education in Nigeria is gradually becoming an increasing enterprise judging by the rise of new private institutions being license.
- The impact of education on the rise in unemployment is however negligible, else unemployment should be on the decrease in the country. Nigeria being a developing country follows the educational systems established since the colonial era. This brings about a dearth in local content and application to the economic and economic development in Nigeria.

Twaalfhoven, W. (2004) put it in perspective, that: Globalization, the rapid development of technology and the lower cost of travel have completely changed the nature of work. Students, as an essential part of the future work force, have to deal with an increasingly complex and uncertain world due to profound economic, social and technical structural change. In dictators of these changes are shifts to service and knowledge-based economies, the rise of emerging economies as well as societal challenges such as environmental sustainability and aging populations. In addition, the explosive growth of social networks demonstrates that boundaries are less easy to define. It is a challenge for higher education institutions to prepare students for work in a dynamic, rapid ly changing entrepreneurial and global environment.

In this context there is a specific opportunity for high-growth entrepreneurship at higher education institutions. Universities, especially technical universities, can be seen as engines of scientific invention and technological development. Invention and technological development can be transformed into innovation. Entrepreneurship is important as a diffusion mechanism to transform scientific inventions into new product and service innovations. Universities play key roles in promoting the talents of students, graduates and researchers. What distinguishes institutions of higher education from other institutions in society is their role in creating knowledge and producing high-potential graduates and researchers. For entrepreneurship education, focusing on institutions of higher education offers the chance to develop knowledge intensive high-growth enterprises from all academic disciplines, not just technical ones. Higher education institutions should create an environment that fosters entrepreneurial mind-sets, skills and behaviours across their organizations. Universities can teach students how to start and grow enterprises in ways that benefit society.

Technical universities in particular provide potential breeding grounds for high technology and high-growth companies or gazelles, therefore universities play a pivotal role in the development of entrepreneurship in Nigeria, while acting as a resource hub, connecting researchers, students, entrepreneurs, companies and other stakeholders. It is important to involve stakeholders inside and outside of higher education institutions. For example, entrepreneurs and entrepreneurial leaders acting as role models for students should be an essential part of entrepreneurship modules and programmes. If the young people and students are to enter the business world and entrepreneurship it is necessary to involve business people and entrepreneurs in the academic education process. Academic start-ups have the potential to grow faster than others, universities can obviously offer support in entrepreneurship education for high growth. In this context it is important to boost regional business potential and activities and to promote international networking and cooperation. Moreover, it is important to offer students techniques that can be applied in the real world. Therefore, a shift from classical models of teaching to experiential learning approaches is essential. Entrepreneurship education at universities can be regarded as theoretically based real life experience. High-growth entrepreneurship can be seen as an adequate form of education for developing high potential students and graduates that can become future opinion leaders and perhaps role models.

Chapter 5

Worldview Issues of Small and Medium Enterprises

5.1. Worldview on the Issue of Small and Medium Enterprises

Research on successful and unsuccessful new and young growing enterprises can aid understanding of growth processes and barriers to growth. Research on how the growth of new enterprises can be better supported, particularly through education at colleges and universities also would be useful. Hence, all over the world increasing attention is being paid to the potential of university education to facilitate high growth enterprises. For example, research has demonstrated that a high-growth entrepreneur in Europe is better educated than other entrepreneurs and the general population. In Europe, most founders of technology based enterprises have a university degree. Research carried out in Germany has shown that enterprises started by individuals with university degrees tend to grow faster than enterprises founded by n on -academics (Egeln, 2000). (Ley, 2006) states that, Research is also needed on how to motivate and nurture the entrepreneurial potential of female students who traditionally may be less inclined to found and manage innovation-oriented high-growth firms. Another group of specific interest is ethnic and immigrant entrepreneurs, who – though often not innovative in the beginning – may introduce novel business practices and subsequent product and service innovations within established communities.

Universal antecedents of start-up behaviour displayed by different groups of society and success factors for high-growth enterprises are hard to determine due to the heterogeneity of technology sectors and individual development paths. The strength of entrepreneurship education however, is to influence people's attitudes towards entrepreneurship and the prospects and feasibility of becoming a growth entrepreneur. Entrepreneurial activities of university students depend to a large extent on perceived barriers to and support for new venture creation. "A perceived lack of relevant experience and a lack of self-confidence" are two reasons often cited by students and new graduates for not engaging in entrepreneurship after graduation (European Commission, 2008; also see the recent empirical studies by Linan, 2008).

The perception of graduates as to whether founding one's own business is desirable personally and socially also impacts entrepreneurial activity (Krueger 2000). Generally, it is necessary to sensitize students to entrepreneurial thinking and taking action in the right direction. One focus can be to sensitize students that creating a new venture can be an alternative to employment. It is important to raise awareness and generate motivation for the discipline of entrepreneurship. Strategically, two target groups may be addressed; first, entrepreneurship education in "a wider sense" and entrepreneurship education in "a narrower sense" (Koch, 2003).

The former entails offering courses to students who will be involved in catalyzing entrepreneurship in their future employments. Raising the awareness and understanding of the specific needs of country and being able to step up to it profitably in different sectors (for example, venture capital and market acceptance of product innovation) will be the essential catalyst here. Entrepreneurship education in the narrower sense follows a direct approach, developing students' competences and entrepreneurial intentions towards starting a business as a career option. In particular, interested students can be offered suitable modules or events for qualifying in different entrepreneurial fields of competence in order to deepen their knowledge in the further course of their studies. With this in view, it is essential to train students in the skills they will need to develop the entrepreneurial ability of creating business ideas, identifying and recognizing opportunities, setting up a business and managing its growth. Students must be prepared "for a life world of much greater uncertainty and complexity involving frequent occupational, job and contract status change, working in a world of fluid organizational structures, greater probability of self- employment and wider responsibilities in family and social life" (Gibb, Hannon 2006). A supportive learning environment for entrepreneurial education is essential.

Business formations do not regularly take place directly after the completion of one's studies. Graduates often prefer to gain practical experience before they start their own businesses. The systematic setup of alumni organizations to facilitate sustained contact with graduates is of special importance; in particular to perpetuate awareness that entrepreneurship is a viable career option. A particular challenge will be to address the substantial opportunity costs of becoming an entrepreneur within the group of high potential university graduates with excellent employment prospects (Wiklund, *et al.*, 2004).

5.2 Models of the Employment Relationship

Scholars conceptualize the employment relationship in various ways. A key assumption is the extent to which the employment relationship necessarily includes conflicts of interests between employers and employees, and the form of such conflicts. In economic theorizing, the labour market mediates all such conflicts such that employers and employees who enter into an employment relationship are assumed to find this arrangement in their own self-interest. In human resource management theorizing, employers and employees are assumed to have shared interests (or a unity of interests, hence the label "unitarism"). Any conflicts that exist are seen as a manifestation of poor human resource management policies or interpersonal clashes such as personality conflicts, both of which can and should be managed away. From the perspective of pluralist industrial relations, the employment relationship is characterized by a plurality of stakeholders with legitimate interests (hence the label "pluralism), and some conflicts of interests are seen as inherent in the employment relationship (e.g., wages v. profits). Lastly, the critical paradigm emphasizes antagonistic conflicts of interests between various groups (e.g., the competing capitalist and working classes in a Marxist framework) that are part of a deeper social conflict of unequal power relations. As a result, there are four common models of employment:

1. *Mainstream economics:* employment is seen as a mutually advantageous transaction in a free market between self-interested legal and economic equals.
2. *Human resource management (unitarism):* employment is a long-term partnership of employees and employers with common interests.
3. *Pluralist industrial relations:* employment is a bargained exchange between stakeholders with some common and some competing economic interests and unequal bargaining power due to imperfect labor markets.
4. *Critical industrial relations:* employment is an unequal power relation between competing groups that is embedded in and inseparable from systemic inequalities throughout the socio-politico-economic system.

These models are important because they help reveal why individuals hold differing perspectives on human resource management policies, labor unions, and employment regulation. For example, human resource management policies are seen as dictated by the market in the first view, as essential mechanisms for aligning the interests of employees and employers and thereby creating profitable companies in the second view, as insufficient for looking out for workers' interests in the third view, and as manipulative managerial tools for shaping the ideology and structure of the workplace in the fourth view.

5.2.1. Globalization and employment relations

The balance of economic efficiency and social equity is the ultimate debate in the field of employment relations. By meeting the needs of the employer; generating profits to establish and maintain economic efficiency; whilst maintaining a balance with the employee and creating social equity that benefits the worker so that he/she can fund and enjoy healthy living; proves to be a continuous revolving issue in westernized societies.

Globalization has effected these issues by creating certain economic factors that disallow or allow various employment issues. Economist Edward Lee (1996) studies the effects of globalization and summarizes the four major points of concern that affect employment relations:

1. International competition, from the newly industrialized countries, will cause unemployment growth and increased wage disparity for unskilled workers in industrialized countries. Imports from low-wage countries exert pressure on the manufacturing sector in industrialized countries and foreign direct investment (FDI) is attracted away from the industrialized nations, towards low-waged countries.
2. Economic liberalization will result in unemployment and wage inequality in developing countries. This happens as job losses in un-competitive industries outstrip job opportunities in new industries.
3. Workers will be forced to accept worsening wages and conditions, as a global labour market results in a “race to the bottom”. Increased international competition creates a pressure to reduce the wages and conditions of workers.
4. Globalization reduces the autonomy of the nation state. Capital is increasingly mobile and the ability of the state to regulate economic activity is reduced.

5.2.2. Employment fraud/modern slavery

Employment as a contract can be oral or written. Most employment agencies prefer the written format of contract to seal openings for incurring suit made against them. It is also an opportunity for the agencies to set their terms and conditions that the potential employee would ascribe, failure of which, they risk termination of the employment and / or pay. In other instances, the contract can be made oral as is the case with most casual labours and family related businesses. Sad to mention, it is on this contractual basis that there are open grounds for fraud and exploitation. Employers may terminate the oral contract at will as a result making the employees at a disadvantaged position for a legal suit.

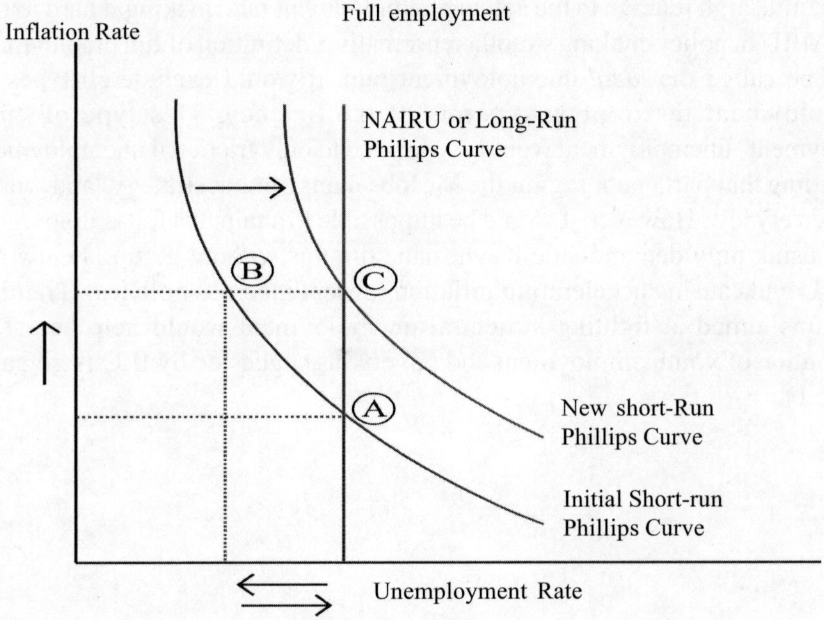

Short-Run Phillips Curve before and after Expansionary Policy, with Long-Run Phillips Curve (NAIRU). *Note:* however, that the unemployment rate is an inaccurate predictor of inflation in the long term.

In demand-based theory, it is possible to abolish cyclical unemployment by increasing the aggregate demand for products and workers. However, eventually the economy hits an "inflation barrier" imposed by the four other kinds of unemployment to the extent that they exist. Historical experience suggests that low unemployment affects inflation in the short term but not the long term. In the long term, the velocity of money supply measures such as the MZM ("money zero maturity," representing cash and equivalent demand deposits) velocity is far more predictive of inflation than low unemployment.Some demand theory economists see the inflation barrier as corresponding to the natural rate of unemployment. The "natural" rate of unemployment is defined as the rate of unemployment that exists when the labour market is in equilibrium and there is pressure for neither rising inflation rates nor falling inflation rates. An alternative technical term for this rate is the NAIRU or the Non-Accelerating Inflation Rate of Unemployment. No matter what its name, demand theory holds that this means that if the unemployment rate gets "too low," inflation will accelerate in the absence of wage and price controls (incomes policies).

One of the major problems with the NAIRU theory is that no one knows exactly what the NAIRU is (while it clearly changes over time). The margin of error can be quite high relative to the actual unemployment rate, making it hard to use the NAIRU in policy-making. Another, normative, definition of full employment might be called the *ideal* unemployment rate. It would exclude all types of unemployment that represent forms of inefficiency. This type of "full employment" unemployment would correspond to only frictional unemployment (excluding that part encouraging the McJobs management strategy) and would thus be very low. However, it would be impossible to attain this full-employment target using only demand-side Keynesian stimulus without getting below the NAIRU and causing accelerating inflation (absent incomes policies). Training programs aimed at fighting structural unemployment would help here.The distribution of youth employment and poverty as suggested by ILO is given in (Table 1).

Table 1: Youth Employment, Poverty and Net Rates in selected developing economies, by sex (%)

Country	Year		Women population ratio			Men population ratio	
Benin	2003	58.2	38.0	20.1	43.5	43.6	14.9
Bhutan	2003	58.4	27.2	15.7	51.2	30.1	6.2
Bolivia	2002	41.8	17.9	20.7	60.2	19.6	5.0
Burkina Faso	2003	73.5	47.1	16.1	79.1	53.2	6.1
Cambodia	2004	76.1	37.8	13.4	78.6	37.8	6.7
Colombia	2003	35.4	13.2	43.2	55.6	16.3	26.2
Congo	2005	27.7	53.4	32.0	26.5	52.2	25.2
Gabon	2005	12.3	4.2	30.9	18.1	4.3	20.9
Guatemala	2000	41.1	11.2	44.4	79.8	10.9	6.0
Indonesia	2002	32.0	26.9	40.8	50.6	27.5	19.9
Jordan	2002	6.2	0.9	44.1	32.2	0.9	21.4
Kenya	2005	33.4	17.6	26.2	37.1	18.0	15.9
Liberia	2007	55.8	81.3	22.9	55.4	82.5	14.6
Mali	2006	55.6	45.0	29.6	66.2	48.8	11.3
Mexico	2004	31.4	1.3	37.4	59.3	1.5	10.4
Mozambique	2003	68.0	68.8	11.5	52.3	65.1	9.4
Nepal	2003	90.1	51.4	5.5	76.5	47.1	7.4
Nicaragua	2005	29.5	12.5	38.7	70.3	16.5	10.3
Niger	2005	22.8	58.5	67.9	61.4	61.1	21.3
Pakistan	2005	15.0	19.9	66.4	62.3	17.5	10.9
Thailand	2002	42.5	0.6	12.6	51.8	0.6	8.5
Togo	2006	63.6	32.3	13.3	59.0	38.5	6.3
Uganda	2005	65.0	42.8	10.7	65.4	43.6	4.6
Viet Nam	2006	52.3	19.5	11.0	55.5	18.2	8.3
Nigeria	2013	42.1	35.8	16.8	58.9	57.19	56.3

Source: Adapted from ILO (employment) and World Bank (poverty) calculations based on national household surveys.

Chapter 6

Issues in Growth and Development in Nigeria: Economic and Non-Economic

6.1. Issues in Growth and Development in Nigeria

Nigeria has attempted through various reform programmes to better the lots of her citizens by providing dividend of development, such as education, health, social infrastructure. Broadly speaking, the features of an underdeveloped economy create obstacles in the way of economic development, and hamper economic progress. These features emerge out of economic, social, political, religious and institutional factors. It would be wrong however to conclude that only economic factors are responsible for poverty or economic backwardness of a country. Non-economic factors are equally responsible for the under development of an economy. The factors discouraging economic development may be classified into economic and non-economic factors which are:

Economic Factors	Non-Economic Factors
1. Vicious Circle of Poverty	1. Underdeveloped Human Resources
2. Deficiency of Capital	2. Political Instability
3. Market Imperfections	3. Socio-cultural constrains
4. International Forces	4. Religious factors
5. Low Agricultural Productivity	
6. Lack of Entrepreneurs	

Available evidence from other countries seems to suggest to some extent that economic growth lead to economic development when properly administered. In Nigerian today it is apt to ask: **Why is economic growth not leading to development?** With respect to the current situation in Nigeria, I will briefly discuss the following issues relevant to development among several others.

6.2. Corruption and Governance Challenge in Nigeria

The most basic requirement for faster rate of development is improved governance, stability and peace. Some observers have noted that absence of these virtues is inimical to sustainable development. Nigeria is currently being affected by these problems to the extent that in large areas of the nation real development is impeded. The causes/reasons for this situation varied across the nation but the underlining causes could be traced to poverty, under-development, and lack of economic diversification and marginalization of the population. Unless these issues are redress, development however perceived will be difficult to achieve despite high GDP. Corruption is a major obstacle to meaningful development in Nigeria. This fact is acknowledged locally and internationally. Government has however put in place institutions to combat this menace, but we all know that the problem is far from over.

Recently, the World Bank in its report, as reported in Punch Newspaper (August, 10, 2012) noted that" investors in Nigeria lost 10% of their revenue to poor quality infrastructure, crime, insecurity and corruption in 2011". The 2011 Corruption Perception Index (CPI) published by Transparency International ranked Nigeria as 143 out of 183 surveyed countries. The CPI for Nigeria is 2.2 (with 10.0 being a clean country and being the most corrupt). The government and in particular, the people of Nigeria must agree to put a stop to corrupt practices as government alone is not capable without the cooperation of the populace. The current report of foreign firms relocating to nearby nations is a clear reflection of the negative impact of corruption and insecurity on development.

In a remark on Institutional and Governance Reforms in Nigeria by the The Brookings Institution (2007), it noted with concern the effect of corruption on the growth and development of the nation. The report posited thus: "a bane of Nigeria's existence since the oil boom of the 1970s has been the reputation for corruption, largely justified, but also partly the result of perception. Corruption and poor governance affected growth and public service delivery in Nigeria in various ways. Corruption distorts the climate for doing business and serves as a tax on private investments. In a corrupt environment, resources for human capital and other needed investments, such as in infrastructure, health and education are often diverted. There are various ways in which this may occur, including procurement fraud, patronage for access to services, absenteeism and misuse of facilities. In particular, poor households are disproportionately hurt in communities where corruption is most prevalent. A key tenet of the reform program, therefore, was to prioritize anti-corruption measures as a central component of the reforms."

Effort at combating corruption in Nigeria is an on-going but complex problem. Corruption has devastating effect on the nation, reduces our citizens to poverty and deprived the nation of any sustainable development. As a result of massive corruption, Nigeria faces many challenges in sustaining economic growth and improving its broad development indicators.

6.3. Addressing Poverty and Inequality

The current Nigerian Poverty Profile Report of the Nigerian Bureau of Statistic 2010 (NBS) reveals a disappointing scenario. The following measurements are used to evaluate poverty in Nigeria:

a) Relative Poverty (RP) is defined by reference to the living standard of majority of a given society. In 2004, the Nigerian RP measurement was 54.4% but increased to 69.0%, or 112.518 million, in 2010.

b) Absolute Poverty (AP) is defined in term of the minimal requirement necessary to afford minimal standard of food, shelter, clothing etc. 54.7% Nigerians were living in poverty in 2004 but this increased to 60.9% or 99.28 million Nigerians.

c) The Dollar-per-day measurement refers to the proportion of those living on less than $1 per day poverty line. With respect to this measurememt, 51.6% Nigerians were living below US$1 per day in 2004 but increased to 61.2%.

Income inequality, measured by Gini coefficient rose from 0.429 in 2004 to 0.447 in 2010 indicating greater income inequality during the period. Forecast by NBS for 2011 is not cheering as it estimate that poverty, using relative, absolute and dollar-per-day may further increase to about 71.5%, 61.9% and 62.8% respectively. It is an unexplained paradox, a self–contradictory statement, that despite the fact that Nigerian economy is said to be growing, the proportion of people living in poverty increases yearly, because economic growth is not development and economic growth has not been successfully translated to development through appropriate government policies. Consequently, he number of poor people continues to increase despite several development and poverty reduction programmes and rural citizens constitute the majority of the population and poverty is deemed to be acute in the rural relative to the urban sectors. The number of poor people will continue to increase except correct national steps are put in place involving the opinions and consultation with poor rural dwellers. In effect, poverty assessment and reduction must be participatory.

Nigeria is a large country with higher concentration of rural dwellers. The rural dwellers are about 80% of the population. Economic development in Nigeria is unlikely to succeed unless the issue of (rural) poverty is addressed. The problems associated with poverty are enormous. They include issues such as: education, health, land tenure, credit, production and marketing processes.

The causes of poverty in Nigeria are very complex but the basic causes are closely related to problems of access and endowment. Specifically, they are according to the World Bank (1985);

- Inadequate access to employment opportunity for the poor;
- Inadequate asset such as land and capital for the poor;
- Inadequate access to the means of authority in poor regions;
- Inadequate access to market for goods and services which the poor can sell;
- Inadequate access to education, health, and water services;
- Inadequate involvement of the poor in the design of development programmes;

Regarding the characteristics of poverty, it can be safely assumed that:

i. While rural poverty dominates in Nigeria, urban poverty is a growing and explosive problem propelled by employment crises especially of the youth.

ii. Gender inequalities are a key feature of poverty pattern.

The character of poverty as described above seems to suggest that the poverty reduction strategy will include:

a. Labour intensive growth;

b. Agricultural growth and provision of basic health services, education and infrastructure in particular to rural areas;

c. Investment in human resources, encouragement of growth of investment and employment in urban areas with the focus on the informal sector.

6.4. Building Human Resources

Building human resources involve investing in people, through education and health. Education, including skills acquisition is the basic foundation for development. A literate and skilled society is better organized, productive, active and development oriented. Also investing in people is essential for growth, development and poverty reduction. Developed countries have invested massively and continuously on human resources development. Given the present situation, Nigeria needs to invest on its human resources development through adequate education of its youth, via skill acquisition and formal education of its people, especially women. Goal 3 of the MDG is to Promote Gender Equality and Empower Women. Recent evaluation however shows that a gradual improvement in the proportion of girls enrolled in primary school, though noteworthy, is not yet enough to meet the target. There are still fewer girls than boys in school. There are signs of backsliding in the number of girls in tertiary education. Measures to encourage girls to attend school, particularly by addressing cultural barriers in the north of the country, and to provide the economic incentives for boys to attend school in the south-east, are urgently required. Confronting regional variations in the determinants of gender inequality requires policies based on an understanding of the underlying socioeconomic, social and cultural factors. State and local government efforts will thus be critical to the achievement of this goal, UNDP (2007).

The government has been implementing education reforms since 2006 to improve access to education, especially with the introduction of Universal Basic Education. The government increased the share of education in budget spending from 4% in 2010 to 6% in 2011. There have been some positive results. The literacy rate for the 15-24 age groups was 80% in 2008, up from 64.1% in 2000. Education quality is, however, generally low and varies considerably across the country. The West African Examination Council said that in 2010 only 25% of Nigerian candidates obtained credits in English language, mathematics, and at least three other subjects (the minimum requirement to enter a tertiary institution). Social indicators on health and education remain weak. The 2011 UN Human Development Report ranked Nigeria 156th out of 187 countries. In 2010, Nigeria was 142nd out of 169 countries. The government has sought to improve the health care system. Health spending in the federal budget increased from 4% in 2010 to 6% in 2011. Also, the Federal Government has adopted several policy initiatives to strengthen the health system, including a National Strategic Health Development (NSHD) plan. Despite these efforts, health outcomes remain poor. According to the United Nations 2010 MDG Report, the infant mortality rate was 75 deaths/1000 live births, the under-five mortality rate was 157/1000, the maternal mortality rate was 545 deaths per 100 000 live births and the proportion of births attended by skilled personnel was 39%.

6.5. Making Agricultural Development a Priority in Nigeria

Despite oil's revenue dominance, agriculture plays a significant role in the Nigerian economy, accounting for 34.5% of GDP in the first quarter of 2012 according to the National Bureau of Statistics. Sustainable expansion of agriculture should play a key role in unleashing inclusive economic growth, reducing poverty and enhancing food security. The government's Agricultural

Transformation Agenda targets these objectives. Agriculture's importance to poverty reduction goes far beyond its direct impact on farmers' incomes. There is evidence that increasing agricultural productivity has benefited millions through higher incomes, more plentiful and cheaper food, and by generating patterns of development that are employment-intensive and benefit both rural and urban areas. More importantly, it has provided the spur to economic development outside agriculture where growth and job creation are faster and wages higher. Rapid increases in agricultural output, brought about by increasing land and labour productivity, have made food cheaper, benefiting both the urban and rural poor, who spend much of their income on food.

In a special report on Development Outreach (vol.10, no.3, 2008), the editor advocated for making agriculture a priority and called for renewed effort to place agriculture at the center of the development agenda in order to meet the MDG goal of halving extreme poverty and hunger by 2015. As an economic activity agriculture contributes to development as a source of livelihood and food security and provider of environmental services. Agriculture can be

- The main source of growth for national economy,
- Provider of investment opportunities for the private sector,
- A private driver of agri-related industries and rural non-farm economy

Majority of poor people live in the rural area and they live on less than $2 a day and must depend on agriculture. Given where they live and what they do, promoting agriculture is imperative for meeting the MDG of halving poverty and hunger by 2015. And continue to reduce poverty and hunger several decades after. Though agriculture alone will not be enough to massively reduce poverty, but poverty reduction will not happen without agriculture.

Poverty is overwhelmingly a rural problem. That is so, and that it endures is the failure of agriculture to deliver livelihood for the large proportion of our people who depend on it. Agriculture growth can reduce poverty directly, by raising farm income and indirectly, through labour market and by reducing food prices. Increased farm productivity increase farm income and participation of smallholders in this agricultural growth will obviously benefit them. Agricultural growth also reduces poverty via labor market because agricultural growth is capable of creating employment opportunities for the poor. Lastly, ability to increase non-tradable staple foods will reduce domestic food prices, especially to the poor, in addition to the urban dwellers, who spend a substantial amount of their income on food.

Agriculture has a special poverty reducing power. China, India and Ghana, for example, have all had relatively high agricultural growth rates and reduced poverty substantially as a result. As noted by Development Outreach Ghana's growth and poverty reduction over the last 15 years is Africa's success story. Real GDP has grown at more than four percent a year since 1980 and at more than five percent since 2001. The poverty rate fell from 52% in 1991 to 29% in 2005, making Ghana the only country in sub-Saharan Africa to have already met the MDG goal of halving poverty.

Similarly, IFAD (2005) in its publication noted "that reversing recent disappointing trends in agriculture's performance is critical if poor countries are to escape the trap of slow growth and poverty. This is particularly true in sub-Saharan Africa, where growth in agricultural output has barely kept pace with population. Productivity has stagnated, slowing wider economic growth and exacerbating poverty with it.

Table 1: GDP per capita of African countries

African Rank	World Rank	Nation	GDP per capita ($)	Total GDP World Rank
1	32	Eq. Guinea	30,233	110
2	35	Seychelles	27,008	172
3	56	Libya	17,665	-
4	61	Botswana	16321	117
5	62	Gabon	16,086	106
6	77	South Africa	11,440	28
7	89	Tunisia	9,795	78
8	95	Algeria	8,515	48
9	96	Namibia	7,565	123
15	136	Nigeria	2,661	38

Source: World Bank (2014)

Although, from table 2, Nigeria is not in the same league as those below the per capita income of $2,000. Nigeria's GDP per capita at PPP is the lowest at $2,661. It occupies the 15th position in Africa but 136st position in the World; this is not laudable at all. Also, looking at the real GDP growth shown in table 3, Nigeria is certainly not the star amongst the countries shown.

Table 2: Real GDP Growth

	2007	2008	2009	2010	2011	2012	2013	2014
Botswana	2.6	4.8	8.3	8.9	3.7	-7.9	8.1	8.0
Eq. Guinea	3.8	9.7	1.3	21.4	10.7	4.6	-0.5	7.7
Gambia	7.0	-0.9	1.1	3.6	5.7	6.4	6.5	-4.4
Ghana	5.6	5.9	6.4	6.5	8.4	4.0	8.0	14.4
Kenya	5.1	5.9	6.3	7.0	1.5	2.7	5.8	4.4
Mozambique	7.9	8.4	8.7	7.3	6.8	6.3	6.8	7.3
Nigeria	10.5	6.5	6.0	6.4	6.0	7.0	8.0	7.4
Uganda	5.8	10.0	7.1	8.1	10.4	4.1	6.2	5.9
South Africa	4.6	5.3	5.6	5.5	3.6	-1.5	3.1	3.5

Source: www.africaneconomicoutlook.org.

Table 3: Human Development Index in selected African countries for the year 2014

Country	Human Development Index (HDI)	World Rank
Botswana	0.634	119
Ghana	0.558	135
Nigeria	0.471	153
Tanzania	0.476	152
South Africa	0.629	121

Source: World Bank (2014)

We have drawn out the statistics for five countries. It is obvious from table 6.3 that poverty is greater in Nigeria than the other countries shown. Turning to other statistics, the one that is used is the Human Development Index (HDI) as presented in table 4. It measures life expectancy, literacy, education and standard of living for countries worldwide. It is often referred to as a standard means of measuring well-being. The rank of Nigeria was 153 in 2012 out of 187 countries! We are still very far behind using all these indicators.

Table 4: Poverty in Selected African Countries for the year 2014

Country	Population below $1 a day (%)	World Rank
Botswana	30.3	67
Ghana	28.6	78
Nigeria	70	6
Tanzania	36	52
South Africa	50	24

Source: www.indexmundi.com/g/r.aspx?v=69

Chapter 7

Causes and Consequences of Poverty

7.1. Rural Poverty in Nigeria

The Federal Republic of Nigeria has a population of more than 160 million – the largest in Africa – and a fast-growing economy. Agriculture is the mainstay of the economy, contributing about 40 per cent of GDP. The agriculture sector employs approximately two-thirds of the country's total labour force and provides a livelihood for about 90 per cent of the rural population. Nigeria is the world's largest producer of cassava, yam and cowpea – all staple foods in sub-Saharan Africa. It is also a major producer of fish. Yet it is a food-deficit nation and imports large amounts of grain, livestock products and fish. Nigeria's huge agricultural resource base offers great potential for growth. Recent government policies have started to show results: The agricultural sector reportedly grew by 7 per cent a year between 2003 and 2007, and at a slightly lower rate in recent years. Still, the area of land under cultivation could be doubled. Of an estimated 71 million hectares of arable land, only about half is presently under production. And there is substantial scope for an increase in irrigation, which now covers only 7 per cent of irrigable land. Irrigation and other inputs would substantially increase average yields for major staple crops, which are below those in other developing countries.

Despite Nigeria's plentiful agricultural resources and oil wealth, poverty is widespread in the country and has increased since the late 1990s. Some 70 per cent of Nigerians live on less than US$1.25 a day.Poverty is especially severe in rural areas, where up to 80 per cent of the population lives below the poverty line, and social services and infrastructure are limited. The country's poor rural women and men depend on agriculture for food and income. About 90 per cent of Nigeria's food is produced by small-scale farmers who cultivate small plots of land and depend on rainfall rather than irrigation systems.The poorest groups eke out a subsistence living but often go short of food, particularly during the pre-harvest period. The productivity of the rural population is also hindered by ill health, particularly HIV/AIDS, tuberculosis and malaria.

Women play a major role in the production, processing and marketing of food crops. Yet women and households headed solely by women are often the most chronically poor members of rural communities. Men have higher social status and, as a result, more access to schooling and training. In recent decades, the number of men migrating from rural areas in search of employment has increased, and the number of households headed solely by women has grown substantially. Rural infrastructure in Nigeria has long been neglected. Investments in health, education and water supply have been focused largely on the cities. As a result, the rural population has extremely limited access to services such as schools and health centres, and about half of the population lacks access to safe drinking water. Neglect of rural infrastructure affects the profitability of agricultural production. The lack of rural roads impedes the marketing of agricultural commodities, prevents farmers from selling their produce at reasonable prices, and leads to spoilage. Limited accessibility cuts small-scale farmers off from sources of inputs, equipment and new technology, and this keeps yields low.

As the population swells and puts pressure on diminishing resources, escalating environmental problems further threaten food production. Land degradation as a result of extensive agriculture, deforestation and overgrazing are already severe in many parts of the country. Drought has become common in the north, and erosion caused by heavy rains, floods and oil pollution is a major problem in the south and south-east. Civil unrest also aggravates poverty. Religious and ethnic tensions continue to brew in different parts of Nigeria, erupting into outbreaks of violence and leading, in turn, to escalating poverty and malnutrition.

7.2. Causes and Consequences of Poverty Today

According to the World Bank, more than one billion people today live on less than $ 1 per day. About 70% of those people are women, and almost half of the population of sub-Saharan Africa survives at that income level. We know that every 3.5 seconds, a child dies in the developing world from poverty-related circumstances. At endPoverty.org we are deeply committed to serving and empowering the very poor to lift themselves out of poverty. But what is poverty really? While on the surface poverty is often defined as a lack of income or assets, in the day-to-day lives of the very poor, poverty becomes a network of disadvantages, each one exacerbating the others. The result is generation after generation of people who lack access to education, health care, adequate housing, proper sanitation and good nutrition. They are the most vulnerable to disasters, armed conflict and systems of political and economic oppression and they are powerless to improve their circumstances. These conditions often carry with them dysfunctional family and societal relationships, paralyzingly low self-esteem, and spiritual darkness. Poverty is a lack of hope. It's clear that handouts and traditional aid are not enough to solve the problem of poverty and its many entanglements. End Poverty.org seeks to equip the poor to free themselves from poverty in a holistic way.

7.2.1. Measuring poverty

A common method used to measure poverty is based on incomes or consumption levels. A person is considered poor if his or her consumption or income level falls below some minimum level necessary to meet basic needs. This minimum level is usually called the "poverty line". What is necessary to satisfy basic needs varies across time and societies. Therefore, poverty lines vary in time and place, and each country uses lines which are appropriate to its level of development, societal norms and values. Information on consumption and income is obtained through sample surveys, with which households are asked to answer detailed questions on their spending habits and sources of income. Such surveys are conducted more or less regularly in most countries. These sample survey data collection methods are increasingly being complemented by participatory methods, where people are asked what their basic needs are and what poverty means for them. Interestingly, new research shows a high degree of concordance between poverty lines based on objective and subjective assessments of needs.

When estimating poverty worldwide, the same reference poverty line has to be used, and expressed in a common unit across countries. Therefore, for the purpose of global aggregation and comparison, the World Bank uses reference lines set at $ 1 and $ 2 per day (more precisely $ 1.08 and $ 2.15 in 1993 Purchasing Power Parity terms). It has been estimated that in 2001, 1.1 billion people had consumption levels below $ 1 a day and 2.7 billion lived on less than $ 2 a day. These figures are lower than earlier estimates, indicating that some progress has taken place, but they still remain too high in terms of human suffering, and much more remains to be done.

7.3. Why Poverty Persist in Nigeria

The following are some of the factors why poverty persists in the country:

1. *Poor macro-economic and monetary policies* – These have contributed to high poverty rate. The Nigerian Naira that used to exchange for 1.20 Naira to one British Pound in 1986, now exchanges for 240 Naira. The outcome has made it almost impossible for small and medium enterprises (SMEs) to thrive, which would have contributed to the reduction of poverty through job creation.

2. *Globalization* – This means trade liberalization. Nations are supposed to trade freely amongst themselves for the mutual benefits of their people.

As for Nigeria, there are only two major goods that could be marketed – crude oil and agricultural produce. The latter is totally neglected with the discovery of the former in a commercial quantity. The gains of globalization as currently being enjoyed by developed nations seem to be eluding Nigeria because of her inability to export manufactured goods abroad like other developed nations.

Good Governance - Nigeria is yet to experience good governance, this has deprived her people the dividends of democracy. Policies of government are still being formulated without due consideration to the welfare of citizens, the outcome of which is further poverty. At this juncture, the little efforts that have been made by the government to control unemployment to an extent shall be examined. Records from the Central Bank of Nigeria (1987), show the following efforts made by the government:

1. Youths Empowerment/ employment and skills programme. This was done in terms of provided minor jobs or 'white-colar' jobs.
2. Agricultural programmes such as the 'Operation Feed the Nation' by the Obasanjo Military Administration.
3. Provision of small scale industries and graduates employment.
4. Special public works (which were temporary). Some of these efforts are still in place while others are forgotten. For example, on the pages of the Businessday Newspaper, on August 4th (pg 13), 12th (pg 49), 21st and 27th, 2009 were the following headings revealed
   - Intels' running of Onne Port Complex creates 8,000 jobs. (NB:Intels is Integrated Logistics Services, a private company in Port Harcourt working with some oil Companies).
   - Lagos Government creates 156,000 jobs in 6 months. (NB: The same government also provides temporary job opportunities with manage. able earnings to secondary school students during long vacations as well as vocational and skill training).
   - Edo State Government lifts embargo on employment. Adding 1 additional to the grade 16 ceiling, making it 17 for employment opportunities.
   - IITA creates 6,500 jobs through cassava project. (NB: IITA means International Institute of Tropical Agriculture).

The Federal Government too is not left out. It creates jobs for graduates and provides vocational and skill training for them. Sectors such as communication, transportation, oil, etc both private and public are not left out in jobs creation.

With these facts above, one is likely to think that enough has been done. What? With more than 140 million populace and over 50 million labour forces consisting of greater percentage of the unemployed, a discouraging figure? Much still needed to be accomplished by the government and all concerned individuals to provide jobs for the millions of Nigerian unemployed. And these provisions must spread across the nation, reaching both rural and urban areas.

Chapter 8

Obstacles of Nigeria in Economic Growth

8.1. Major Obstacle to Nigeria Economic Growth

In Nigeria, high transport prices are rightly seen as a major obstacle to economic growth of the region. This has been documented in several studies. One of them (Amjadi and Yeats, 1995) even concluded that, in Nigeria, transport costs are a higher trade barrier than import tariffs and trade restrictions. In this paper, it is analyzed, identified and quantified the factors behind Nigeria's high prices of road transport, with the purpose of helping policy-makers adopt appropriate actions to reduce transport prices to domestic and international trade. Before the presentation of the results, a distinction has to be made between logistics costs, transport prices, transport costs and vehicle operating costs in order to avoid the usual confusion. This distinction is useful because prices may or may not reflect transport cost:

- Transport prices are the rates charged by a transport company or a freight forwarder to the shipper or importer. Transport prices usually are the result of negotiated rates between the shipper and the transport service provider. Transport prices normally cover transport costs, the operator's overheads and profit margin.
- Transport costs are the costs the transport operator incurs when transporting a cargo. In addition to VOC, transport cost includes other indirect costs, such as license fee, roadblocks, etc.
- Vehicle operating costs (VOC) include the various direct costs the transport provider must pay to operate a given vehicle, notably labor, capital, fuel, tyres, maintenance and depreciation cost of a vehicle.

However, a number of studies that looked at the determinants of the high transport costs in Africa did not arrive at a wide consensus, but rather identified different factors as being the main contributors to the high prices. The factors included:

low productivity of the trucking industry in Africa, notably due to infrastructure constraints (Pedersen, 2001), low levels of competition between service providers (Rizet and Hine, 1993), or weak infrastructure (Limao and Venables, 2001). This latter study also indicated the poor quality of infrastructure as accounting for most of Africa's poor trade performance. From a cross country regression, the study concluded that trade was highly sensitive to transport costs – a 10 percent drop in transport costs increases trade by 25 percent, and transport costs are sensitive to the quality of infrastructure, as measured by such variables as the density of the road and rail network. In the past, large investments in improving road infrastructure were predicated on the basis that transport prices would be reduced. The World Bank since the 1970s have actively supported the transport corridors in Africa and during many years such support focused almost exclusively on improving infrastructure. While the improvements were essential to facilitate road transport and resulted in lower costs for the trucks carrying cargo on the corridors, no clear impact on the transport prices was evident. Thus, the end-users of road transport services did not seem to fully benefit from the lower transport costs and better service quality that the improved infrastructure could possibly allow. Since many research attempts in the past did not arrive at a conclusive result, the current study has made a deliberate attempt to expand both the breadth and depth of the research and can claim to be original in the following areas:

- It is the first comprehensive and practical effort in the last fifteen years to measure and quantify the high transport prices in Nigeria.
- It is the first attempt of this kind in Nigeria and elsewhere in the world to disaggregate and clearly define the various input factors into three tiers of cost and prices, i.e. vehicle operating costs, transport costs (incurred to commercial transport providers) and transport prices (incurred to end users), and try to understand how each of them effect the final transport prices.
- It is also the first study that clearly recognizes the diversification in transportation market across Africa continent with an attempt to measure with the same yard stick the costs, prices and performances of the transportation industry across the four sub-regions of Africa. Contrary to what most literature has presented, transport costs are not excessively high in Africa but transport prices are, especially in Central Africa mainly due to official and unofficial market regulation and structure in trucking services.

8.1.1. Transport costs and rates

Transport systems face requirements to increase their capacity and to reduce the costs of movements. All users (e.g. individuals, enterprises, institutions, governments, etc.) have to negotiate or bid for the transfer of goods, people, information and capital because supplies, distribution systems, tariffs, salaries, locations, marketing techniques as well as fuel costs are changing constantly. There are also costs involved in gathering information, negotiating, and enforcing contracts and transactions, which are often referred as the cost of doing business. Trade involves transactions costs that all agents attempt to reduce since transaction costs account for a growing share of the resources consumed by the economy. Frequently, enterprises and individuals must take decisions about how to route passengers or freight through the transport system. This choice has been considerably expanded in the context of the production of lighter and high value consuming goods, such as electronics, and less bulky production techniques. It is not uncommon for transport costs to account for 10% of the total cost of a product. This share also roughly applies to personal mobility where households spend about 10% of their income for transportation, including the automobile which has a complex cost structure. Thus, the choice of a transportation mode to route people and freight within origins and destinations becomes important and depends on a number of factors such as the nature of the goods, the available infrastructures, origins and destinations, technology, and particularly their respective distances. Jointly, they define transportation costs.

Transport costs are a monetary measure of what the transport provider must pay to produce transportation services. They come as fixed (infrastructure) and variable (operating) costs, depending on a variety of conditions related to geography, infrastructure, administrative barriers, energy, and on how passengers and freight are carried. Three major components, related to transactions, shipments and the friction of distance, impact on transport costs. Transport costs have significant impacts on the structure of economic activities as well as on international trade. Empirical evidence underlines that raising transport costs by 10% reduces trade volumes by more than 20%. In a competitive environment where transportation is a service that can be bided on, transport costs are influenced by the respective rates of transport companies, the portion of the transport costs charged to users.

Rates are the price of transportation services paid by their users. They are the negotiated monetary cost of moving a passenger or a unit of freight between a specific origin and destination. Rates are often visible to the consumers since transport providers must provide this information to secure transactions. They may not necessarily express the real transport costs. The difference between costs and rates either results in a loss or a profit from the service provider. Considering the components of transport costs previously discussed, rate setting is a complex undertaking subject to constant change. For public transit, rates are often fixed and the result of a political decision where a share of the total costs is subsidized by the society. The goal is to provide an affordable mobility to the largest possible segment of the population even if this implies a recurring deficit (public transit systems rarely make any profit). It is thus common for public transit systems to have rates that are lower than costs. For freight transportation and many forms of passenger transportation (e.g. air transportation) rates are subject to a competitive pressure. This means that the rate will be adjusted according to the demand and the supply. They either reflect costs directly involved with shipping (cost-of-service) or are determined by the value of the commodity (value-of-service). Since many actors involved in freight transportation are private rates tend to vary, often significantly, but profitability is paramount.

8.1.2. Costs and time components

Transportation offers a spectrum of costs and level of services. Among the most significant conditions affecting transport costs and thus transport rates are:

- **Geography**: Its impacts mainly involve distance and accessibility. Distance is commonly the most basic condition affecting transport costs. The more it is difficult to trade space for a cost, the more the friction of distance is important. It can be expressed in terms of length, time, economic costs or the amount of energy used. It varies greatly according to the type of transportation mode involved and the efficiency of specific transport routes. Landlocked countries tend to have higher transport costs, often twice as much, as they do not have direct access to maritime transportation.
- **Type of product**: Many products require packaging, special handling, are bulky or perishable. Coal is obviously a commodity that is easier to transport than fruits or fresh flowers as it requires rudimentary storage facilities and can be transshipped using rudimentary equipment. Insurance costs are also to be considered and are commonly a function of the value to weight ratio and the risk associated with the movement. As such, different economic sectors incur different transport costs as they each have their own transport intensity. With containerization the type of product plays little in the transport cost since rates are set per container, but products still need to be loaded or unloaded from the container. For passengers,

comfort and amenities must be provided, especially if long distance travel is involved.

- **Economies of scale**: Another condition affecting transport costs is related to **economies of scale or the possibilities to apply them** as the larger the quantities transported, the lower the unit cost. Bulk commodities such as energy (coal, oil), minerals and grains are highly suitable to obtain lower unit transport costs if they are transported in large quantities. For instance, moving a barrel of oil over 4,000 km would cost $1 on a 150,000 deadweight tons tanker ship and $3 on a 50,000 deadweight tons tanker ship. A similar trend also applies to container shipping with larger containerships involving lower unit costs.

- **Energy**: Transport activities are large consumers of energy, especially oil. About 60% of all the global oil consumption is attributed to transport activities. Transport typically account for about 25% of all the energy consumption of an economy. The costs of several energy intensive transport modes, such as air transport, are particularly susceptible to fluctuations in energy prices.

- **Trade imbalances**: Imbalances between imports and exports have impacts on transport costs. This is especially the case for container transportation since trade imbalances imply the repositioning of empty containers that have to be taken into account in the total transport costs. Consequently, if a trade balance is strongly negative (more imports than exports), transport costs for imports tend to be higher than for exports. Significant transport rate imbalances have emerged along major trade routes. The same condition applies at the national and local levels where freight flows are often unidirectional, implying empty movements.

- **Infrastructures**: The efficiency and capacity of transport modes and terminals has a direct impact on transport costs. Poor infrastructures imply higher transport costs, delays and negative economic consequences. More developed transport systems tend to have lower transport costs since they are more reliable and can handle more movements.

- **Mode**: Different modes are characterized by different transport costs, since each has its own capacity limitations and operational conditions. When two or more modes are directly competing for the same market, the outcome often results in lower transport costs. Containerized transportationpermitted a significant reduction in freight transport rates around the world.

- **Competition and regulation**: Concerns the complex competitive and regulatory environment in which transportation takes place. Transport services taking place over highly competitive segments tend to be of lower cost than on segments with limited competition (oligopoly or monopoly). International competition has favored concentration in many segments of the transport industry, namely maritime and air modes. Regulations, such as tariffs, cabotage laws, labor, security and safety impose additional transport costs, particularly in developing countries.
- **Surcharges**: Refer to an array of fees, often set in an arbitrary fashion, to reflect temporary conditions that may impact on costs assumed by the transporter. The most common are fuel surcharges, security fees, geopolitical risk premiums and additional baggage fees.

The transport time component is also an important consideration as it is associated with the service factor of transportation. They include the transport time, the order time, the timing, the punctuality and the frequency. For instance, a maritime shipper may offer a container transport service between a number of North American and Pacific Asian ports. It may take 12 days to service two ports across the Pacific (transport time) and a port call is done every two days (frequency). In order to secure a slot on a ship, a freight forwarder must call at least five days in advance (order time). For a specific port terminal, a ship arrives at 8AM and leaves at 5PM (timing) with the average delay being two hours (punctuality).

Mobility tends to be influenced by transport costs. Empirical evidence for passenger vehicle use underlines the relationship between annual vehicle mileage and fuel costs, implying the higher fuel costs are, the lower the mileage. At the international level, doubling of transport costs can reduce trade flows by more than 80%. The more affordable mobility is, the more frequent the movements and the more likely they will take place over longer distances. A wide variety of transport costs can be considered.

Freight on Board (FOB): Is a transport rate where the price of a good is the combination of the factory costs and the shipping costs from the factory to the consumer. In the case of FOB, the consumer pays for the freight transport costs. Consequently, the price of a commodity will vary according to transportation costs and distance.

Costs-Insurance-Freight (CIF): Is a transport rate that considers the price of the good, insurance costs and transport costs. It implies a uniform delivered price for all customers everywhere, with no spatially variable shipping price. The average shipping price is built into the price of a good. The CIF cost structure can be expanded to include several rate zones, such as one for local, another for the nation and another for exports.

Terminal costs: Costs that are related to the loading, transshipment and unloading. Two major terminal costs can be considered; loading and unloading at the origin and destination, which are unavoidable, and intermediate (transshipment) costs that can be avoided.

Linehaul costs: Costs that are a function of the distance over which a unit of freight or passenger is carried. Weight is also a cost function when freight is involved. They include labor and fuel and commonly exclude transshipment costs.

Capital costs: Costs applying to the physical assets of transportation mainly infrastructures, terminals and vehicles. They include the purchase or major enhancement of fixed assets, which can often be a one-time event. Since physical assets tend to depreciate over time, capital investments are required on a regular basis for maintenance.

Transport providers make a variety of decisions based on their cost structure, a function of all the above types of transport costs. The role of transport companies has sensibly increased in the general context of the global commercial geography. However, the nature of this role is changing as a result of a general reduction of transport costs but growing infrastructure costs, mainly due to greater flows and competition for land. Each transport sector must consider variations in the importance of different transport costs. While operating costs are high for air transport, terminal costs are significant for maritime transport. Several indexes, such as the Baltic Dry Index, have been developed to convey a pricing mechanism useful for planning and decision making. Technological changes and their associated decline in transport costs have weakened the links transport modes and their terminals. There is less emphasis on heavy industries and more importance given to manufacturing and transport services (e.g. warehousing and distribution). Indeed, new functions are being grafted to transport activities that are henceforward facilitating **logistics** and manufacturing processes. Relations between terminal operators and carriers have thus become crucial notably in containerized traffic. They are needed to overcome the physical and time constraints of transshipment, notably at ports.

The requirements of international trade gave rise to the development of **specialized and intermediary firms** providing transport services. These are firms that do not physically transport the goods, but are required to facilitate the grouping, storage and handling of freight as well as the complex paperwork and financial and legal transactions involved in international trade. Examples included freight forwarders, customs brokers, warehousing, insurance agents and banking, etc. Recently, there has been a trend to consolidate these different intermediate functions, and a growing proportion of global trade is now being organized by multi-national corporations that are offering door to door logistics services.

8.2. Monetary Policy Impacts on Growth, Inflation and Exchange Rates

To utilize this framework we used data going back as far the 1980s. This is to enable us understand the interaction of key variables, rather than assessing the impacts of changes in monetary policy alone. Here, we were interested in looking at the medium term impact of changes in monetary policy on key variables such as inflation, economic growth and exchange rates. The first VAR model that was used analyzed the following variables treasury bill rates, inflation, exchange rate changes and real economic growth. The second model substituted the rate of growth of the money supply (M2 growth) for the treasury bill rate and retained the other endogenous variables. Variable definitions are presented in Table 1 below.

Table 1: Variable Definitions and sources

Variables	Definitions	Sources
Interest Rate	90-Day interest rate: %	Central Bank of Nigeria
Inflation	4-quarter rate of change of consumer price index: %	Central Bank of Nigeria
Growth	4-quarter rate of growth of real GDP: %	Central Bank of Nigeria
Exchange Rate	4-quarter rate of change of exchange rate: %	Central Bank of Nigeria
Price of Oil	Natural log of the $ price of oil	Central Bank of Nigeria
Money Growth	Rate of growth of M2 Supply: %	Central Bank of Nigeria

Source: Adapted from CBN Report, 2014.

Diskey, using the adjusted fuller tests. In order to check for significance, the endogenous variables were subjected to tests. Pair-wise Granger causality tests was also carried out on all the variables in order to determine which of these relationships are Granger "caused" at standard significant level. The Granger - causality tests was conducted for these relationships. In order to better assess the underlying macro-economic dynamics, we built two simple VAR models and used each of these to describe monetary policy experiments.

Chapter 9

Issues and Challenges for Growth in Nigeria

9.1. Challenges for Nigeria

Nigeria, at independence in 1960, was largely a producer and net exporter of primary products. The six major agricultural products then were cocoa, rubber, palm oil, groundnut, cotton and palm kernel. Although there existed mining and quarrying activities, these were of negligible percentage and never counted for the economy as a whole. In other words, agricultural produce and raw materials constituted the sole foreign exchange earner for the country. Specifically, the Nigerian state as an exporter of agricultural goods had 69.4% of its total GDP for the year 1963/64 comprising the six aforementioned agricultural commodities (Olaloku, 1979:8).

The trend of having agriculture as the main foreign exchange earner for Nigeria stopped in the early 1970s when the country was suddenly awash with petrol dollars arising from the quadruple increase in the price of oil in the world market. From 1972 onwards, oil gained ascendancy over all other commodities as the largest contributor to the GDP, and also as a major foreign exchange earner. There was substantial increase in oil production accompanied by a sharp increase in the global market price of high-grade crude oil from low price of $3.8 per barrel in October 1973 to a skyrocketing price of $14.7 per barrel in January 1974. This trend continued till 1981 when the price of crude oil attained a high level of $38.77 per barrel. Within the same period, total revenue from oil rose correspondingly from ₦ 1 billion to ₦ 4 billion while external reserves increased from ₦ 180 million to ₦ 3.7 billion in 1975 (Osaghae, 1998:96).

The increase in national wealth resulted in the government of the day embarking on rapid expansion of the public sector and squandering of the wealth on expanding distributive instead of productive capacity; and on increased dependence on external goods and inputs (Ibid). By 1978, oil (Petroleum and associated components) composed a total of 89.1% of Nigeria's export as against agriculture, which had plummeted in its contribution to export to 6.8% in the same year. The fact of the matter is that Nigeria's commodity pattern has, since the advent of oil, been a mono-cultural one with the product being the only one the country depends upon for its foreign exchange earning - a situation that has constrained the pace of its developmental efforts. Thus, it can be unequivocally asserted that Nigeria's development impotence cannot be solely attributed to colonialism, neo-colonialism and imperialism as often perceptively advanced by scholars of radical persuasion; but also to its own weak domestic economic structure. There is therefore the necessity for the country to diversify its economic base in order to confront the challenges of contemporary globalization process and remain relevant in the scheme of world events. The economies of South East Asian Countries especially the Asian Tigers namely Taiwan, Hong Kong, Singapore and South Korea as well as Thailand, Malaysia and Vietnam are successful stories of how diversification has enhanced their economic developments. The success recorded by this Newly Industrialized Countries (NIC) can be duplicated by Nigeria in its current efforts toward economic transformation.

For instance, as a means of liberalizing its economy, revitalizing its financial sector and generating foreign exchange earnings, Thailand placed top priority on exports. Its economy was liberalized thus encouraging foreign investors. The country's manufacturing sector has emerged as a formidable engine of development in the kingdom with traditional exports in the agricultural sector performing quite successfully. In 1996, Thailand exported about US$9billion worth of goods to the European Union alone, and by 1997, the country's exports had become highly competitively priced for foreign buyers (Akinbobola, 1999:65). Similarly, in the manufacturing sector, there has been a phenomenal increase in the exports categories. A dramatic development has been the automotive industry in which several big automobile makers have set up production lines in Thailand, thus making the country to become a regional leader in the production and exportation of automotive. In 1996, automotive exports to the European Union alone amounted to US$205million signifying an increase of 38% over the preceding year (The Economist, cited by Akinbobola: 66).

It is interesting to note however that in spite of Thailand's advent of high technology, the country has remained deeply committed to its agricultural base and has indeed emerged as one of the worlds leaders in agro business. In general, Thailand has confronted the challenge of globalization by liberalizing its economy, embracing internationalization of capital and opening new markets across the length and breadth of the world. Hence,Thais products (manufacturing and agricultural) are found all over American, European, Asian and African countries (Ibid). Nigeria can tremendously benefit from Thailand's experience. The Nigerian economy must be completely diversified rather than remaining mono-cultured.

Furthermore, there is the need to ensure a free competitive economy. Taiwan represents one of those Asian nations that have transited from government guidance to a free competitive economy, and thus liberalizing its economy for foreign investors. Today, there exists in Taiwan, foreign investments in construction, power generation, oil refining, real estate, telecommunications and gas stations. In concrete terms, foreign investment has been quite staggering. For instance, United States accumulated investment in Taiwan amount to US $ 7 billion representing about 30% of total approved foreign investment in the country. Taiwan has huge foreign investments in South East Asian countries and all over the globe. The country also has a large chunk of human resources ready to be utilized by foreign firms. All these have been made possible because of the country's dynamic efforts towards a free competitive economic structure.

The challenges posed by globalization for Nigeria are multifarious. These are considered in the same manner that we articulated the forces that propel it. First, for the country to be fully integrated into the world economy and in order to harness the benefits of such integration, it must embark on serious technological revolution. As Shamsudden Usman pungently remarks, the pervasiveness of technology is such that a country ignores it at its own peril (Usman, Op cit: 56). In order to ensure technological revolution, the country's technological base must be developed. Science and technology should be made a formidable part of the key strategic area of its developmental effort. The need to fundamentally transform the nation's educational, health, agriculture and industrial development has therefore become an imperative demand. Due attention must equally be focused on information technology and telecommunication. In contemporary global economy, information technology and the ability to use it are very critical. While computers and other electronics gadgets are already in place and do not need to be re-invented, there is the need to ensure that the enabling environment for their uses is provided. Hence, adequate basic infrastructures such as power supply and telecommunications must exist regularly and uninterruptedly. Similarly, industrialization, including manufacturing and fabrication should be re-invigorated, while agriculture should be re-positioned with mechanized farming fully embraced to enhance agricultural productivity.

The second challenge relates to economic liberalization. The liberalization policy of the Asian Tigers tremendously enhanced their development. This entails liberalizing the economy, internationalizing of capital, opening new markets and attracting new investments. This poses a great challenge to Nigeria. We agree with Usman that the Nigerian economy must not only be diversified, but also built on sound economic policies including those that will necessarily ensure increased domestic savings; continued reform of the domestic financial sector; opening up to foreign capital inflows, while simultaneously protecting the country from the huge destabilizing effects of short-term, speculative capital inflows; and together with other developing countries, continue to champion the necessity for the reform of the global financial system that ensures shared prosperity and a greater inducement to the development of the weaker countries (Ibid:58).The heavy dependence of the country on crude oil exports has unprecedentedly exposed the economy to the boom-and-bust cycles and the concomitant unstable and unpredictable volume of revenue receivable by the government (Yaqub, 2003:41). For instance, Nigeria's total export receipts from goods, services and transfers dropped from ₦ 10,899.6 million in 1979 to ₦ 7,884.2 million in 1983. The countrys import, at the same time increased from ₦ 9,890.1 million to ₦ 11.022.1million during the same period (CBN Economic and Financial Review 1981-1986). This also affects contributions to the country's Gross Domestic Product (GDP). For instance as at 1999, oil and gas contributed 36.5% of the GDP, while agriculture (including livestock) accounted for 32.8%. In the same period, wholesale and retail trade contributed 16.6%, while manufacturing accounted for only 5.5% of the GDP (Anya, 2001:15). There is therefore the need for the country to diversify the economy and focus on non-oil sector, particularly agricultural and mineral resources.

In addition, the country should maintain a healthy investment climate that can be cashed on by foreign investors. It should also pursue efficient and effective economic management of the country's resources so as to raise the people's standard of living and overall economic development of the nation. It must be stressed however that in its bid to liberalize the economy, the unbridled activities of the Multinational Corporations in the country should be closely monitored and controlled so that the country will not be reaped off by foreign capital. The fact that globalization entails opening up does not signify that the economy should be completely left uncontrolled. Democracy constitutes the third challenge that globalization poses for Nigeria. Democracy has become an acceptable form of governance in the world system. Indeed, no authoritarian or dictatorial regime is fashionable any longer in the global environment. As observed by the UNDP National Human Development Report: The end of the cold war provided a historic opportunity for a world-wide liberal democratic revolution. In view of the western industrialized market economies, the collapse of authoritarianism and socialist central planning has revalidated the claim of liberal democracy, the doctrine of individual freedom and popular democracy, as an ideology of potentially universal validity (UNDP, National Human Development Report for

Nigeria, 2000/2001, p.2). Nigeria regained its civil rule in 1999 after about fifteen years of uninterrupted military dictatorship. The persuasive lack of democratic system of governance for several years had significantly deteriorated the developmental pace of the country. Following the restoration of civilian government therefore, there is the need for democracy to be fully entrenched and sustained. As stressed in the National Human Development Report, there must be a strong and vibrant civil society, good governance, effective dynamic relationship between people and the government and entrusting the management of the state in the hands of men and women of high technical competence and integrity (Ibid, p. 42). The country must be truly committed to, and demonstrate the capacity for good governance as a bedrock for a durable democracy.

In this book, it has been demonstrated that globalization as a phenomenon has assumed a new phase in contemporary global political economy, and that Nigeria must equip and package itself effectively to confront its challenges in the 21 st century. The paper also articulates the view that globalization in its current dimension is characterized by three distinct but interrelated features: the integration of the economic systems of nation states into global economy; the primacy and supremacy of international markets and international competitiveness; and the phenomenal rise in the internationalization of labour, capital and portfolio investments. The book critically examined the three forces that propelled globalization namely technological revolution, economic liberalization and democratic governmental system. It submits that the new thrust of globalization makes it imperative for nation states to understand the intricacies of the phenomenon so as to enable them devise strategic ways to harness its advantages.The challenges of globalization must be faced by developing nations for them to be relevant in the global scheme of things. For Nigeria, in particular, the challenges are enormous but not insurmountable.

Nigeria cannot afford to remain aloof and be completely isolated from the global economic interplay. The country should make conscious effort to transform the economy so as to fully harness the benefits of globalization. The various endogenous and exogenous constraints that have continued to undermine macro economic stability in the country should be seriously tackled. The critical sectors of the economy that constitute the basis for sustainable growth e.g. agriculture, mining, industry and energy should be the core concern of economic policy, and hence properly harnessed for the countrys development. The nation's technological base should be re-activated; hence the need to earnestly complete the Ajaokuta steel rolling mill, the Aladja steel plant and the Katsina, Jos and Osogbo steel rolling mills. In concert with other African states and using the instrumentality of the New Partnership for African Development (NEPAD), Nigeria should work assiduously to ensure the collaborative involvement of all stakeholders in African development through the promotion of regional economic blocs. Finally, it must be stressed that given the nature and character of the Nigerian state with its inherent weak domestic base, globalization, certainly has its adverse implications on the nation's economy. However, in order to minimize the negative effects and harness the whatever benefits of the current growing and overwhelming globalization trend; there is the necessity for a complete restructuring and transformation of the Nigerian political economy.

9.2. Barriers of SMEs Trading

He subject of trade barriers has been a recurring issue in the country for years. Thus, a forum that emphasised the need for African countries, West African countries in particular, to trade with each other was organised Thursday by the Lagos Chamber of Commerce and Industry (LCCI). The event, tagged: 'Stakeholders' Forum on Breaking Trade Barriers in West African sub-region,' drew participants from across the country. Joseph Annan, deputy minister of trade and industry of Ghana, gave the keynote address, while Lewis Olu Ogunojemite, comptroller of Customs, presented a paper at the forum, and recalled Hillary Clinton, US secretary of state, calling on African countries to increase trade with each other, in other to realise the enormous potential that exists within the continent.The customs boss argued that although ECOWAS have made strides towards free trade, "full market integration remains an ambition. The efforts to eliminate tariffs have begun and all members, except Liberia, have eliminated tariffs on unprocessed products. Finished products also only attract the Export Trade Liberalisation Scheme (ETLS) tax; revitalisation of trade liberalisation has begun with the fast track initiative between Ghana and Nigeria; efforts are on going to remove non-tariff barriers, however, progress on removal of these barriers is harder to assess, as data on such barriers are inadequate and by their very nature, they are not directly measurable."

For him, there are other underlying structural constraints such as deep-seated distrust among member states, which fundamentally limits the depth and progress in regional integration, there are systemic problems that hamper the development of national economies and invariably impede the regional integration, there is the reluctance to adhere to integration programmes due to concerns over losses and uneven gains, insufficient analytical and technical support also limit implementation of some integration instruments, for instance, the trade liberalisation scheme.In terms of progress made in advancing the goals of regional integration, Ogunojemite said the results are mixed with a general sense that there remain substantial gaps in achievements."Some encouraging advances have been made in the area of information and communication technology (ICT), due to global advances in this sector, as well as the increased role of the private sector in enhancing service provision. However, in comparison to other parts of the world, the sub-region is still a long way from realising universal access to ICT," he argued.

Quoting World Bank (2003), Hester Hopkins, general manager, operations, COTECTNA, said trade can be a powerful force for growth and poverty reduction. Countries that have increased the share of trade in their GDP have grown faster and reduced poverty more rapidly. She stressed the need for trade facilitation in the sub-region to reduce the complexity and cost of the trade transaction process, and ensure that all these activities take place in an efficient, transparent and predictable manner.Those she listed as important players in the process of trade facilitation include: government and government agencies, transporters/logistic providers, financiers/insurers, and traders.For Hopkins, trade facilitation reduces transaction costs, allows for faster delivery/clearance, predictable trade rules, increased foreign investments, reduced corruption, increased revenue and increased compliance. For her too, trade facilitation principles include harmonisation of applicable laws and regulations, simplification, that is the use of ICT to exchange information efficiently, standardisation which has to do with administrative and commercial formalities, procedures and documents; transparency, that ensures that information, requirements are processed for crossing borders, clear, specific and easily accessible for all involved.

Citing the SWEPRO/National Board of Trade, Sweden, 2002, she argued that trade facilitation results in direct benefits both to governments and to the business community in terms of increased economic efficiency, better security, faster delivery of goods, and reduced costs.

Hopkins argued that the World Bank observes that the top 10 trade facilitation economies, which include Singapore, Hong Kong, China and the Republic of Korea, share four characteristics as follows:

- Introduced paper-free electronic data interchange systems and pre-arrival submission of customs declarations.
- Make use of a risk management inspection system, where less than 10 percent of containers are physically inspected.
- Have systems allowing authorised persons, who meet prescribed criteria to benefit from simplified procedures.
- Have excellent port infrastructure.

Roberts Orya, managing director/CEO, Nigeria Import Export Bank, also reeled our data on the economy of the sub-region and place of Nigeria in the economy."According to IMF/World Bank data, the ECOWAS region attained a total GDP of $565 billion and a population of 289 million in 2010. Of this figure, Nigeria contributed over 60 percent, with a GDP of $374 billion and a population of 150 million.The Economic Community of West African States (ECOWAS) Treaty, which is a regional grouping of 15 member countries, is expected to eliminate all tariffs, restrictions and quotas towards facilitating free movement of people, goods and services, thereby promoting social and economic harmonisation, deepening trade and encouraging large-scale investment.The above mentioned goals are expected to achieve the attainment of a free trade area, a unified custom union/establishment of a common external tariff, creation of a common market, and the creation of an economic union (harmonisation of economic, agricultural, industrial and monetary policies, etc.Orya further said: "Intra-regional trade remains a relatively small part of the ECOWAS economic activity, accounting for about 10 percent of the total trade of the member states. In 2008, the total ECOWAS intra-regional trade was valued at $ 6.9 billion, while total ECOWAS trade with the world was valued at $64.4 billion." Historically, the value of ECOWAS intra-regional trade has over the years been increasing from $ 3.2 billion in 2002 to peak at $ 9.9 billion in 2006 before it nose-dived to $ 6.6 billion and $ 6.9 billion in 2007 and 2008 respectively. "In tonnage, the volume had similarly increased from 4.2 million tons in 1996 to 13.4 million tons in 2008."

For Orya, "With regards to Nigeria's level of ECOWAS intra-regional trade, the aggregate non-oil exports stood at $189.96 million in 2009 declining from $228.81 in 2008, and representing 9.7 percent of Nigeria's non oil export in 2009 account to the 2009 CBN Annual Account."

The low level of ECOWAS intra-regional trade, according to Orya, are largely traced to some tariff and non-tariff barriers, some of which include lack of appropriate/adequate infrastructure at the region that is characterised by lack of poor rail systems, roads, energy etc. This is more compelling when viewed against the empirical evidence that cargo volumes have leaped about 300 per cent from 4.7 million tons in 1998 to 13.4 million tons in 2008 without corresponding increase/improvement on basic logistical infrastructure; barriers to movement of goods and services, due to multiple check points/border posts with attendant consequences of time and cost overruns and eventual uncompetitive pricing of goods at market destination. Femi Deru, president, LCCI, said trade within the ECOWAS sub-region has become more imperative today than ever before. "Global competitiveness is a major factor in achieving efficiency, which is in itself dependent on market size and the ability to enjoy economies of scale," he said. For him, with a robust market of about 300 million people, significant benefits of economies of scale would be enjoyed by West African firms in the event of full market integration.

Deru argued there is a strong relationship between the development of intra regional trade and the promotion of economic integration in the sub-region. "Integration is in fact the main vehicle for the boosting of trade within the sub-region, with integration, African economies would be stronger and their capacity to cope with the challenges of globalisation would be enhanced. Africa would also be better positioned to respond to the challenges of peace and security in the continent," he said. For him, some progress has been made under the current integration initiatives in Africa, but on the whole, the performance in economic integration has not been quite satisfactory. Joseph Annan, who delivered the keynote address, addressed the issue of constraints to West African trade listing high transport cost and poor infrastructure and border measures and regulatory environment. He argued it is well known that transportation costs are extremely high between our countries. He said road freight tariffs in sub-Saharan Africa are in the range of $ 0.04-0.14 per ton per kilometre compared to $ 0.01-0.04 per ton per kilometre in other developing regions. "Trade to landlocked countries in particular is especially costly, with import costs three to five times the global average. It takes 116 days to move a container from Bangui, Central African Republic to the nearest port," he said. For him, time-consuming and costly border measures and unharmonised and cumbersome regulatory systems are even greater barriers to trade than poor physical infrastructure.

Annan, the implementation of existing agreements is the most important step; negotiating new agreements could even be counterproductive. He said since the implementation of regional agreements can only advance at the pace of the slowest moving country, pressure from peers and business leaders is needed to speed this process. He advised that there is need to enforce existing trade protocols especially ETLS and free movement of persons. For him, at the national level, countries need to think carefully about how to promote West African trade, and then act to integrate these strategies into their development plans.

9.3. Foreign Trade Barriers in Nigeria

(a) Nigeria trade summary

The U.S. goods trade deficit with Nigeria was $ 34 billion in 2008, an increase of $ 4 billion from $ 30 billion in 2007. U.S. goods exports in 2008 were $ 4.1 billion, up 47.7 percent from the previous year.

Corresponding U.S. imports from Nigeria were $ 38.1 billion, up 16.2 percent. Nigeria is currently the 44th largest export market for U.S. goods.

The stock of U.S. foreign direct investment (FDI) in Nigeria was $ 190 million in 2006 (latest data available).

(b) Import policies tariffs

In September 2008, the Nigerian government issued the 2008-2012 Common External Tariff (CET) Book that harmonizes its tariffs with its West African neighbors' under the Economic Community of West African States (ECOWAS) Common External Tariff (CET). Nigeria has been partially implementing the CET since 2005. The new tariff regime has five tariff bands and import duties have been reduced on a number of items, such as rice, cigars, and manufactured tobacco. The five CET tariff bands are: zero duty on capital goods, machinery, and essential drugs not produced locally; 5 percent on imported raw materials; 10 percent on intermediate goods; 20 percent on finished goods; and 35 percent on goods in certain sectors. The fifth band is still under negotiation among member countries. Adoption of the CET is part of ongoing economic reforms aimed at improving Nigeria's trade and investment environment and harmonization of economic policies in the sub-region. There remains resistance within the Nigerian government and the private sector to deepen trade reforms.

Companies state that high tariffs, nontransparent valuation procedures, frequent policy changes and unclear interpretations by the Nigerian Customs Service (NCS) continue to make importing difficult and expensive, and often create bottlenecks for commercial activities. Some importers complain that tariffs are excessively high and that the Nigerian government sometimes uses arbitrary reference prices for valuation purposes. This problem is aggravated by Nigeria's dependence on imported raw materials and finished goods and affects both foreign and domestic manufacturers. Reportedly, many importers resort to undervaluing and smuggling to avoid paying full tariffs. Transparent and proper implementation of the CET would be an important step toward resolving most of these problems.

(C) Non tariff measures

Though the government continues to ban certain imports, citing the need to protect local industries, the new tariff book reduces the number of items on the import prohibition list from 44 to 26. Items removed from the list include corn, sorghum, millet, wheat flour, crude vegetable oil, biscuits, sugar confectioneries (including white chocolate), fresh and dried fruit, flowers (both fresh and plastic), toothpaste, envelopes, diaries, greeting cards, exercise books, bentonites, barites, calendars, cutlasses, axes, pick axes, spades, shovels, fully built mudguards, wheel barrows, and electric generating sound proof casings.

Items remaining on the import prohibition list include: bird's eggs, cocoa butter, powder and cakes, pork, beef, live birds, frozen poultry, refined vegetable oil and fats, cassava, bottled water, spaghetti, noodles, Fruit juice in retail packs, nonalcoholic beverages (excluding energy drinks), certain textile products, and bagged cement. Companies were awarded concessions to import bagged cement for a limited time to bridge supply gaps. The cement concessions terminated at the end of December 2008, but the companies have applied for an extension.

(d) Customs administration

Nigeria practices a destination inspection policy for imports. Under this policy, all imports are inspected upon arrival into Nigeria, rather than at the ports of origin. Nigeria port practices continue to present major obstacles to trade. The country's long list of items prohibited for import, coupled with incorrect declaration of goods by importers, result in 95 percent of containers being physically examined. This delays the clearing process and increases costs. Nigeria's uneven application of import and labeling regulations make importing high-value perishable products difficult. Disputes between Nigerian agencies over the interpretation of regulations often cause delays, and frequent changes in customs guidelines slow the movement of goods through Nigerian ports. Importers report erratic application of customs regulations, lengthy clearance procedures, high berthing and unloading costs, and corruption. These factors can contribute to product deterioration and may result in significant losses for importers of perishable goods.

Realizing that delays at the ports significantly increase the cost of doing business in Nigeria, the Nigerian government plans to implement a 48 hour cargo clearance policy at the ports. Roads coming in and out of the ports are decaying, and overuse results in around-the-clock traffic congestion. There is no rail system transporting freight in and out of ports. As such, despite the 48 hour custom clearance policy, congestion leads to ships queuing up to berth at cargo terminals and containers waiting to be transported out of the ports. The bottlenecks resulting from the lack of infrastructure at and around the ports affect the level of efficiency at which goods can be processed for import. Currently over 15 agencies are represented at the ports. In a bid to achieve the 48 hour cargo clearance target at the ports, the Nigerian government plans to withdraw all agencies, except customs, from the ports and improve the technical capacity of customs to handle special cargos through continuous training of personnel. There are also plans to automate all customs payments. Nigeria's government administers various export incentive programs such as tax concessions, export development funds, capital asset depreciation allowances, and foreign currency retention programs in addition to operating Free Trade Zones and Export Processing Zones. According to the 2008-2012 CET Book, most concessions, waivers or exemptions have been stopped; however, the Nigerian Export Promotion Council will continue to implement the Export Expansion Grant scheme to improve non-oil export performance.

Rules concerning sanitary and phytosanitary standards, testing, and labeling are well defined, but bureaucratic hurdles slow the import approval process. Regardless of origin, Nigeria requires that all food, drug, cosmetic, and pesticide imports be accompanied by certificates of analysis from manufacturers and appropriate national authorities; and specified animal products, plants, seeds, and soils must be accompanied by proper inspection certificates. By law, items entering Nigeria must be labeled exclusively in the metric system. U.S. producers and exporters note that relabeling goods to meet this requirement is expensive and limits U.S. exports to Nigeria. The NCS is charged with preventing the entry of products with dual or multiple markings, but such items are often found in Nigerian markets. The National Agency for Food and Drug Administration and Control (NAFDAC) is charged with protecting Nigerian consumers from fraudulent or unhealthy products. The agency continues to focus special attention on eliminating the illicit importation of counterfeit and expired pharmaceuticals, particularly from East and South Asia. NAFDAC's limited capacity for carrying out inspections and testing contributes to what critics have characterized as an occasionally heavy-handed or arbitrary approach to regulatory enforcement which sometimes leads to delays in clearance of legitimate food imports.

Although Nigeria has no laws governing agricultural biotechnology or bio safety, the government is generally supportive of biotechnology. The Federal Ministry of Environment has presented draft Bio safety legislation to the National Council on Environment, the highest decision-making body on environmental issues. If approved, the legislation will be sent to the National Executive Council of Ministers for ratification and then to the National Assembly for its consideration. The draft bill generally portrays products of biotechnology as safe for animal and human consumption; however, it includes a mandatory labeling requirement.

Chapter 10

Challenges and Opportunities in Small and Medium Enterprises

10.1. Challenges and Opportunities in SMEs

The Nigerian transport system faces great challenges and also offers several opportunities. The predisposing factors include:

- *The size of the country:* With area coverage of 923,768.64 km^2, spanning longitude 30E-160E and latitude 40N-140N Nigeria is a comparatively large country. It is inhabited by over 140 million people. Transportation is a crucial instrument for linking the country economically, socially and politically.
- *Nigeria's vast natural resources:* Nigeria is endowed with petroleum, gas and solid mineral resources. The country is also blessed with abundant agricultural and forest resources, which vary by the three ecological zones of the country. The country is endowed with a highly productive open sea with abundant and diverse marine resources within her coastline of 852km bordering the Atlantic Ocean in the Gulf of Guinea and her maritime area of 46,000 km. This variation is a factor in spatial interdependence. Transport plays an important role in the exploitation and distribution of these resources and in the reduction of spatial inequality and in poverty alleviation.
- *The Growth of the Nigerian Economy:* Until the current decade, economic growth posed significant challenges to the Nigerian economy, especially during the decades of the 1980s and 1990s. However, between 1999–2008, the performance of the Nigerian economy improved significantly. Real Gross Domestic Product (GDP) grew at an annual average rate of 5.6% and was the highest in three decades. The economic growth rate of almost two times the estimated 3% growth rate of the population ensured a real per capita output growth of 2.6%. The oil sector tended to constitute a drag on growth. The composition of national output

for the 1999–2008 decade shows that there is still the challenge of achieving broad-based growth. Three large sectors – crop production, oil and gas, and wholesale and retail trade contributed the most to GDP. The role of Manufacturing was limited while wholesale and retail trade, through a high contributor to GDP, had low linkage with domestic production due to low domestic production output and high importation. Overall, the Sectors of the economy are still plagued by numerous problems notably, poor and decaying infrastructure, epileptic power supply, institutions and regulatory deficit, etc. The defective structure of the nation's economy make her heavily dependent on export earnings from crude oil and on import of consumer and capital goods, thereby increasing the country's cargo throughput and boosting the overall maritime trade. The huge volume of cargo generated by Nigeria places her at an economic advantage in South-Saharan Africa.

- *Strategic Location of Nigeria:* by its location, Nigeria has an extensive coast washed by the Atlantic Ocean and a geographically conducive shore which provide the opportunity of port services to landlocked countries in West Africa and particularly Chad and Niger Republic, as well as serve as a hub for transshipment for the West and Central African sub-region. Thus, the demand for transport is high and growing. This National Transport Policy acknowledges the need to support the country's social and economic growth to enable her compete favorably in international markets. This Policy above all, acknowledges the need to develop a transport system that ensures national security, opens up inaccessible areas and serves the overall well being of the people.

10.2. Challenges with Current Financial Structures

There are many challenges because the financial sector in many developing countries are currently facing challenges that imparts need to be explained to put in place employment generating and poverty reducing monetary framework Despite financial liberalization, and in some cases even because of it, financial structures in many developing countries are not fostering development, employment generation and poverty reduction. They also face the following challenges listed below:

- *Real Interest Rates and Interest Rate Spreads are high:*

Real interest rate spreads remain very high in many developing countries, and especially in sub-Saharan Africa. The banking systems are highly concentrated, hold high levels of excess reserves and lend primarily to very large businesses and wealthier consumers. (See Heintz, 2008). Nevertheless, there are examples of well performing financial sectors in some developing countries. Most of these retain a significant degree of regulation (Stallings and Studart, 2006).

- *Credit Creation is too low*

In many poor countries, high spreads and real interest rates are associated with relatively low rates of credit creation.

- *Credit and Capital Flows are: Pro-Cyclical*

When capital flows do come into many developing countries, they are often short-term and pro-cyclical, and sometimes associated with "sudden-stops". These can, in turn, lead to financial crises that can have devastating economic impacts, often especially for women and the poor (Ocampo, 2003).

- *Insufficient Amount of Long-Term, Patient Capital*

The short term and unstable flows of international capital represent a more general, severe problem facing many developing countries — namely, there is a dearth of long-term, patient capital to support long-term investment (Stallings and Studart, 2006). Much capital is of a highly short term nature. As a result, long-term productive investment is extremely difficult and costly to finance.

- *Insufficient Capital for Small and Medium Enterprises and the Poor*

Finally, there is a lack of capital for small and medium enterprises and the poor in most regions of the world Nigeria inclusive (Stallings and Studart, 2006). This problem flows from many of the "stylized facts" described earlier: the high real interest rates and interest rate spreads, the mis-allocation of global financial resources, the short-term and pro-cyclical nature of international capital flows, and the absence of long-term, patient capital. An additional reason is the ability of banks to allocate their assets to high yielding, relatively safe, government securities, which may crowd out lending to SME's and others.

In many countries, the preferred solution to this problem has been a heavy dose of financial liberalization (Nissanke and Aryeetey, 1998; Honohan and Beck, 2007). Recently, liberalization has been combined with the attempt to create a regulatory framework which would monitor and guard against financial instability, while creating space for private financial investment and enterprise, often from abroad. But as the current crisis demonstrates, financial liberalization by itself will not be sufficient to do the job. (Serieux, 2008; Honohan and Beck, 2007; Heintz and Pollin, 2008).

- *Insufficient attention to the problems of rural borrowers*

Even though people living in rural areas make up the vast majority of the world's (and Nigeria's) poor, they are the most underserved in terms of financial services solutions presented below should be designed specifically with the rural poor in mind.

10.3. Central Bank Support for Development Banking and Small-Scale Financial Institutions

The structure of Central Bank support for development Banking and small scale financial institutions may be crucial for a nation's climate. Central Banks have historically played a much larger role in supporting longer-term economic development goals than has become fashionable more recently Epstein (2007a). This has been true both in developing countries and in the now developed countries. Central banks, acting as agents of development, have bought long-term bonds for infrastructure development; kept and administered differential interest rate ceilings in support of favored sectors, such as housing in many developed countries (Epstein, 2007a); used their discount window to discount loans for specialized lending institutions, such as microcredit institutions provided efficient or even subsidized financial services for specialized institutions playing important social roles. These policies have not always been successful, but if planned carefully with monitoring and safeguards, have an important part to play in mobilizing and re-deploying finance for development purposes. For example, Central Bank support for the Grameen Bank in Bangladesh and for Rrural Development Banks in Indonesia have been very successful (Asian Development Bank, 2000). By supporting development banking and small-scale financial institutions, Central Bank will be promoting development that is employment generating nad poverty reducing.

While moderate levels of inflation appear to have no significant costs, using central bank policy to fight inflation does have costs, and some evidence suggests that these costs are felt more by some groups than others. In particular, as shown by Braunstein and Heintz (2006), dis-inflationary monetary policy in developing countries, like that undertaken in connection with inflation-focused monetary policy regimes, has a disproportionately negative employment effect on women, relative to men. That there may be important gender effects of monetary policy, suggesting that this issue needs much more research than it is currently receiving although this is not the main issue in this study. Their empirical exercise explores the effects of inflation reduction on women's and men's formal employment. They compiled data for 51 "inflation reduction episodes" in 17 low- and middle-income countries. 7 to assess the employment effects of inflation reduction periods, they looked at actual employment trends during each inflation reduction episode, disaggregated by gender, and compared these to long-run employment trends. They also examine the indicators that suggest how monetary policy responded during inflation reduction episodes using a similar approach.

In looking at the impacts of disinflation on employment, they distinguished periods when both inflation rates and employment fell, relative to long run trends, from those periods when inflation rates fell but employment expanded. They use the term "contractionary inflation reduction episode" to refer to the inflation reduction periods when employment contracted and "expansionary inflation reduction episode" to those when employment expanded. According to this study:

- If employment contracts during an inflation reduction episode, it is likely that women will experience a larger loss of employment, in percentage terms, than men. In the majority of cases, contractionary inflation reduction has a disproportionately negative impact on women. However, during inflation reduction episodes in which employment expands, the gender-specific impact is ambiguous.
- When Central Banks respond to inflationary pressures by raising real interest rates above the long-run trend they are more likely to experience a slow-down in the growth of employment relative to those countries that keep interest rates in line with or below the long-run trend, with concomitantly higher losses for relative female employment.
- They find that real exchange rates (RERs) seem to impact the gender bias of contractionary inflation reduction episodes. In all cases where women experienced relative employment gains during employment contractions, exchange rates either depreciated or showed no deviation relative to long-run trends.
- Tightening the real money supply also seems to be negatively associated with employment in general and women's employment in particular.

These results suggest that contractionary monetary policy aimed at reducing inflation often has a disproportionately negative impact on women's employment, an effect that may be eased by maintaining a competitive exchange rate. Moreover, inflation reduction episodes where real exchange rates did not appreciate were less unfavorable to women's employment. One of the major socio-economic issues confronting Nigeria is the high and rising number of the unemployed in the country. The problem is even more acute among the youths. The unemployment rate in Nigeria was single digit and on the average less than 4 per cent during the period 1990-1998. The rate jumped from a single digit of 3.5 per cent in 1998 to 17.5 per cent in 1999 (a five hundred per cent increase). Since this period, the unemployment rate has remained double digit. The salient facts about youth unemployment in Nigeria can be summarized as follows:

(a) Out of the total population of 169.15 million in 2012, the unemployed were 18.9 million, of which 11.1 million (58.8 per cent) were youths of between the age of 15 and 34 years;

(b) About 38.7 per cent of the 11.1 million were within the 15 to 24 years age bracket, compared with 20.1 per cent between the age of 25 and 34 years;

(c) 56.8 per cent of the unemployed youths aged 15 and 34 years have less than secondary school education, they are less skilled and inexperienced.

10.4. Trade Policy and the Poverty Reduction in Nigeria

Evaluating trade policy (particularly the 2002 trade policy document in Nigeria) and the extent to which it facilitates or hinders poverty reduction in Nigeria. It examines the state of poverty and the poor in the country and proceeds to evaluate the provisions of the current trade policy document. Based on the intention to provide inputs into the ongoing review of the current trade policy, it also evaluates its consistency with the NEEDS and other domestic development policy documents. Since the last twenty years, economic policy in Nigeria can be characterized by trade liberalization and regional integration which is defined by the radical reducing or removal of trade barriers. The World Trade Organization (WTO) the IMF and the especially the World Bank (WB) have obtained considerable powers to sway policies in countries towards this path. As a part of the global structural adjustment programme, it is assumed and argued that trade liberalization improves the welfare of consumers and trims down poverty. The assertion was two-fold and simple. First, it is argued that liberalization offers wider room for choice from an array of quality goods and cheaper imports also find more lucrative markets in which their products can be sold. A second argument is that, the production of goods in which a country has comparative advantage expands, while the sectors without comparative disadvantage minimize. This is believed to lead to an overall rise in real GDP since there would be reallocation of the productive factors from less efficient sectors to more efficient sectors. 'It is against this setting that many developing countries liberalized their imports, reduced average tariffs and dismantled significant number of Non-Tariff Barriers.Nevertheless the degree of realization of the anticipated rewards of liberalization over the years of agonizing alteration has been the topic of intense debate in many empirical literatures. Some scholars believe these rewards have come, others think it is on the way, while yet another group believe it neither has nor will ever come. Those who give emphasis to trade liberalization also stress the necessity, as a minimum, for an established exchange rate system, cautious monetary and fiscal policies, good governance and a corruption-free Administration of economic policies for trade liberalization to be effective in the long-run. Those against it point to time discrepancy of policy makers, doubts about losses and gains of liberalization processes, increasing barriers to trade by developed countries and unfair competition among other things. The debate rages on and thus far is still far from being decisive on either side. Notwithstanding the fact that the debate is yet to be concluded, policymakers in developing countries are practically being pushed to liberalize. Meanwhile, the gains and costs from the revolutions spun by globalization and integration of world trade are unevenly dispersed and it is not apparent that those who get the best benefits also have the highest cost. Rather to the contrary, evidence show that economies of developing countries often carry the burden of increased openness. The existing Nigerian trade policy document is one of the numerous drawn with strong focus on the global economy, displaying outward orientation in all aspects. Questions thus arise as to the implications of this for poverty reduction which is vital for comprehensive growth and development. It consequently becomes imperative to realize the degree to which trade policy can be used as a successful tool for poverty alleviation. Evaluating the effectiveness of trade policy in promoting livelihood of the poor in Nigeria by evaluating:

- The degree of strengthening or alleviation of the existing marginalization of the poor by trade policy.
- The instruments and provisions within trade to release the poor from the poverty plague.
- How trade can smooth the progress of growth with equity and a better allocation of national wealth for instance, by adding value to Nigeria's primary products such as oil and agriculture products.
- How trade can facilitate the development or attainment of technology and enhancement of local capacities.
- How trade can facilitate investment in human capacity including education, health, food, housing, etc.

The following sections will attempt to show the relationship between international trade and poverty issues, examine the views of operators in the real sector on the impact and trade Policy, Gender and the Livelihoods of the Poor in Nigeria possible directions of trade policy. This goes to contribute to the debate on the impact of international trade on poverty reduction using the experience of Nigeria. The rest of this paper is organized as follows: Chapter two presents a global view of trade, economic growth and poverty focusing on theoretical and empirical links. Chapter three relates the relationship of trade, macroeconomic policy and poverty reduction to Nigeria's experience and history. Chapter four discusses the present trade policies of Nigeria and finally, chapter five summarizes the key emerging issues and recommendations of this paper.

Chapter 11

Approaches for Promotion Programmes in Entrepreneurship

11.1. Support for Start-Up Finance

Access to finance for start-ups is always difficult due to the problem of asymmetric information. Start-ups lack the kind of financial history which banks use to infer the pay back behaviour and the business capacities of the entrepreneur. Innovative start-ups face the additional problem that it is difficult to assess the viability of their business ideas or the prospects of the markets. In addition, banks and other potential investors often decline to get involved when the relatively small amount of capital requested by a new business is out of proportion with the efforts that the investors must make to screen and evaluate the start-up project.

These problems are even greater for companies engaging in IB models, as doing business with the poor is usually riskier – or perceived as more risky – due to BoP customers' low ability to pay and the greater vulnerability of value chains. Support for start-up finance is therefore essential. Local IB start-ups often lack funding at all stages of the start-up process: the idea stage, when business opportunities are identified, the seed stage, when the market is assessed and the business model developed, and the start-up stage, when the company is formally set up and all the necessary equipment is procured. Different kinds of financial support will be necessary at all stages.

(a) For the idea stage, setting up a special fund providing start-up finance for IB models could help motivated entrepreneurs and organisations to undertake the required activities.

(b) Options to address the later stages could include loans from public banks or the use of guarantee schemes to encourage private banks to make loans to IB start-ups. At the same time, support can be given for matchmaking processes through which entrepreneurs are able to network with local investors.Fostering local angel investor networks to promote investment in start-ups and to enable entrepreneurs to identify finance opportunities more easily can also help IB ventures access the financial support they need.

(c) Another field of intervention for PSD projects is cooperation with all stakeholders in order to improve the regulatory environment and infrastructure. Simply improving the factors for 'doing business', such as ease of company registration, contract enforcement, taxes, etc.

(d) Another approach is to encourage local companies with functioning business models – be they SMEs or emerging market multinationals – to adopt inclusive business models by expanding their target groups and adapting their products, services and processes to these new target groups, or by integrating small-scale producers into their value chains. This could be piloted first, starting in limited geographical areas, or with just one specific product or service rather than the full range offered by the company. There are a number of advantages to such a strategy. Unlike start-ups, these companies would be less troubled by problems of access to finance and limited management skills.

11.2. Activities by Entrepreneurer in Small Scale Bussiness

(a) Creating awareness and knowledge about business models

- Programmes can promote awareness and knowledge about IB and IB models by working with business membership organisations and the media, and using role models, competitions and awards.
- When working with business development services in the field of innovation management, programmes can encourage the providers to include modules that foster an enabling culture for inclusive business.

(b) Adapting products and services and stimulating their uptake in the market

- Established local companies might know the local market, but even so, they might still be unfamiliar with the habits and needs of low-income people, especially as they tend to differ between different regions of a country. Companies can benefit from collaboration with local civil society institutions to create products and services that are appropriately adapted to the needs of poor and marginalized people, and to encourage those people to buy the products.

- Together with other local partners such as companies, foundations or public authorities, business programmes can also set up schemes to support partnerships between companies and research institutions or NGOs that engage in joint research & development projects and scale up prototypes developed by grassroots organisations.

(c) Unlocking additional finance

- At some stage, even companies with a sound financial basis will need access to additional funding, in order to expand their target groups, geographical reach or products. Such forums can complement formats like the BoP Sector Dialogue workshops, or they could be implemented in a similar way to the forums organized by New Ventures, which connect local environmental entrepreneurs with investors in Brazil, Colombia, India, Indonesia and Mexico.

(d) Replicate and disseminate

- Perfect the business model by incorporating elements that drive business model success Create more enterprises: build a market through large scale replication of successful, proven models or businesses upgrade sectors already at scale but typically informal and inefficient. Remove barriers or foster conditions to aid enterprises inclusive Enterprise

(e) Measures at policy level

- There are various instruments governments can use at the policy level to support inclusive business models. One area that is often highlighted is statistics. Governments can improve the way they gather statistics in order to provide better market data, particularly with respect to incomes in the informal economy. Up till now, little research has been done on what makes a conducive business environment for businesses in Nigeria

(f) Building inclusive innovation systems

- Business models are based on innovation, which makes innovation policy a relevant policy area for the support of IB models. Policymakers can support inclusive business development by establishing an innovation system that facilitates exchanges and cooperation between academia, the private sector and responsible government actors.

(g) Consumer protection

- Inadequate quality of products and services can be harmful to poor customers, particularly in sectors such as consumer goods or medicine where the drugs they buy might be out-dated or counterfeit. However, it can be a challenge to provide consumer protection in poorer areas where, for instance, many illiterate consumers are unable to read the instructions on medical treatments.
- Vulnerable consumers, such as those at the base of the pyramid, need better information, advice and support services, in order to avoid paying too much for the products and services they receive (poverty penalty). Consumer protection should therefore be supported by an alliance of public and private stakeholders. Civil society organisations can play an important role in tackling unfair practices and poor service delivery to the low-income population.
- Governments can also use appropriate regulation to ensure a conducive context for the fair provision of goods and services to poorer people. It is vitally important to strike the right balance between deregulation in sectors that were once the exclusive responsibility of governments (water, health, education, and energy) and consumer protection. One sector in which such an effort has already been made to strike such a balance is microfinance.

(h) Competition policy

- Opening up these markets for private companies also creates the need for smart competition policy. In the G20 Policy Note on the Business Environment for Inclusive business Models, IFC highlights the need to balance regulation to ensure that well intended policies, such as excessively detailed technical standards and requirements, do not inadvertently stifle the innovation of inclusive business models. Thus, it is very important that governments are aware of the part inclusive businesses can play in solving social and environmental challenges and engage. They should also engage in a strategic public-private dialogue with these businesses, with the aim of developing BoP markets.

(i) Regulatory business environment

- Companies registered as benefit corporations consider their social and environ - mental impacts as well as their financial results when making strategic decisions. Besides giving them a positive public image, this status also means such companies enjoy legal protection when they integrate other (social and environmental) issues into their decision making than purely financial considerations. Another question is whether the local legal and regulatory framework is conducive to micro-franchising. This raises the issue of intellectual property rights, for instance (copyright, protection of brand names, contract design and enforcement, the obligation to disclose all relevant information to franchise holders, registration procedures, etc.).

(j) Fostering networks

- Fostering networks with civil society, with existing regional and global networks, and with institutions. t he programme aims to connect innovators around the country and across regions, to improve the state of learning and support exchanges about sustainable and inclusive business model innovation. G i Z has entered into partnership with D fi D and sankalp, one of India's largest forums, to bring together social enterprises, investors, support providers and government actors. Together they are organizing the capacity building programme for social Enterprises in Bihar the first project of its kind.

(k) Fiscal and other redistributive measures

- Besides improving the enabling environment, both in general and in sector-specific terms, governments can also introduce incentives that increase the supply of, or demand for services and products provided using inclusive business models.
- Governments can create market opportunities for public procurement rules and regulations and setting legal targets for inclusive business procurement. As the largest consumer in any economy, on average the public sector spends between 45% and 65% of its budget on public procurement, which in most cases represents a large proportion of the national GDP. If a greater effort is made to purchase environmentally and socially preferable products and services, the substantial buying power of the public sector might eventually stimulate markets for sustainable products and services, thereby indirectly supporting opportunities for business models.
- There are already many examples in both developed and developing countries of the public procurement process being used to pursue social goals. Switzerland, for instance, requires equal pay for men and women as a condition for public contracts. In India, procurement rules stipulate that certain goods must be purchased from small and micro-enterprises, even if the price is upto 15% higher than that charged by their larger competitors. In Brazil, a law introduced in January 2007 sets out criteria

that are intended to increase the participation of smaller businesses in public procurement. The provision of subsidies is a policy instrument that can be tailored specifically to IB needs.

- Government subsidies are sometimes needed to encourage early entrants to develop BoP markets. They increase the attractiveness of investing in IB ventures by reducing costs and risks, and they can ensure a company's survival during the often critical pioneering phase. Subsidies often also highlight the specific sectors or technologies given political priority, and in this way encourage further investment.

Chapter 12

Small and Medium Entrepreneurship in Africa – A Critical Review

12.1. Practice of SMEs in Africa-Critical Review

Previous studies in Kenya have focused on the role of entrepreneurship education in job creation, enterprise management and economic development (Simiyu, 2010; Mwangi, 2011; Warren, 2011); hardly any have addressed the process of delivering entrepreneurship education. Research provides new knowledge to guide educational practice, by making an initial foray into the teaching of entrepreneurship education in Kenya. The envisaged entrepreneurially empowered TTI graduate is a product of the training function. This book examines factors which affect implementation of entrepreneurship education programme in TTIs in Kenya. The variables investigated were teaching methods, assessment methods, teacher's network, and availability of training resources. The findings show that teachers do not accord entrepreneurship education t he appropriate pedagogical approach that it deserves; they seem resigned to using ineffective traditional pedagogies. Meanwhile, the subject is inadequately resourced by TTI administrations. The results of this research invite the attention of entrepreneurship educators, practitioners and stakeholders to employ effective strategies for dev eloping entrepreneurial key competencies in TTI trainees.

Teachers should abandon their heavy reliance on traditional teaching and assessment methods, and embrace experiential methods. Administrators of TTIs should be sensitized on the importance of providing adequate training resources for entrepreneurship education. A national policy framework is needed to ensure availability of sufficient resources for the programme. Education in Nigeria is devoid of the element crucial to averting the surging rate of unemployment in the country. The provision of entrepreneurial development through education will advance the economy of the nation if much credence would be given to it and ingrained with focus on profitable personal development. Unemployment prevails in the country, hence, the growth of violence, poverty and segregation amongst citizens, because the educational system itself fails to empower the ones passing through it .The core message of the evolving educational policy in Nigeria is devoid of a system of education that emphasizes on the need to culture the country's youth through the knowledge of rudimentary entrepreneurial development, common cultural heritage, and identification of exploitable strengths of structures, systems and cultures of others. If anything, there is a greater need of its augmentation, which can propel the country to work towards enduring peace and harmony.

The research work, therefore, was carried out to examine the strategies of entrepreneurial education carried out in two of the universities pioneering it Federal University of Technology, Akure, and Covenant University Ota, the former being a public university and the latter a private university. The objectives of the study were to appraise if educational styles arouse the interest of students in the industries of their discipline; to explore the effectiveness of entrepreneurial development strategy in education in universities that implement it ; to see if the current university educational system stimulates entrepreneurial creativity in its students The methodology adopted was a mixed analysis of quantitative and qualitative parameters based on the survey design which relied on primary and secondary sources of gathering data , through the use of questionnaires and interview instruments.

Three hundred (300) questionnaires were administered, and two hundred and fifty four (254) were returned. The study adopted quota and simple random sampling technique. The data was analyzed and presented using tables and percentages. Consequently the findings on this research portrayed a huge disparity between the perception and conceptualization of entrepreneurship in the graduates of each university used in the study, creating the significance in the strategies being used to educate the students while in the university. The study shows that entrepreneurial education should be taught in the field and through practical approaches, rather than using theoretical approaches, as the former yields better results for the economy than the latter. It was discovered that graduates from Covenant University have more aptness to creating value, and are self driven to developing some form of business whether or not they have an employment. The study recommends that there should be a working partnership, bridging the gap between the higher institutions and the industry; lecturers should have field experience to aid communication and teaching of the courses. The study also recommends that Universities should work toward becoming entrepreneurial hubs for students and young entrepreneurs; the government should also focus more on the youth age group for entrepreneurship development in the country. Nigeria is bedeviled by a myriad of problems which, despite her oil wealth, inhibit her development and even threaten her continued existence as a sovereign state. Nigeria's socio-political and economic circumstances give the significant indication that many of her problems stem from an origin of artificial colonial construct which lumped together a variety of separate peoples. Fragmentation of the nation is seen as a distinct possibility unless its citizens can be induced to accept a new sense of Nigerian identity, involving a commitment to the survival of the present state as a cohesive entity.

This would necessitate a number of radical changes, not only in the political and economic s tructure of the country but also in the psychology of the people. Nigerians have lived through series of administrations under different governments in Nigeria, and the question still arises, 'is Nigeria a nation at all?', a critical look at what the government calls reform resul ts in personally instituted concept of governance, filling the seats of power with those they believe to be their kin, rather than have professionals in the positions of merit, and a breed of people typified by their integrity of heart, ingrained in the trainings and qualifications they have received in the course of service to the nation. With an increasing number of those who are not gainfully employed or adequately educated in the country, they remain preys as political tools of violence, as it has been seen in the history of violence occurring in the country over a period of time. The country has depended much on oil as its major source of revenue for years, however, the current administration also fails to recognize that the future of the country may very well depend on the economy of its people (the youths), which is possibly the only untapped, ill harnessed, most lucrative resource of the country. If it remains this way in the next ten years, putting into consideration the effect of increased poverty, lack of employment, poor educational system, it is unpredictable what the result will be. It is the result of a sur vey carried out by the National Bureau of Statistics and shows that persons aged 0 -14 years constituted 39.6%, those aged between 1564 (the economically active population), constituted 56.3%, while those aged 65 years and above constituted 4.2%. Analysis of employment dat a for the past 5 years show that the rate of new entrants into the labour market has not been uniform in the past five years. The rate was on the increase from 2007 to 2009 but declined significantly from 2009 to 2010. The rate increased again from 2010 to 2011. Within the five year period there has been an average of about 1.8 million new entrants into the active labour market per year. Evidence from a range of sources including the report of the National Committee on Job Creation and Putting Nigeria to Work (World Bank, 2010) reveals that the biggest drivers of Nigeria's current youth unemployment crises are:

- Lack of jobs
- Lack of skills and experience
- Mismatch between supply and demand for labour
- Numerous barriers to youth entrepreneurship

Plethora of job creation interventions in the public, private and nonprofit sectors, including Nigerian Directorate of Employment (NDE), Industrial Training Fund (ITF), National Poverty Eradication Programme (NAPEP), Small and Medium Enterprises Development Agency of Nigeria (SMEDAN), but reaching fewer than 100,000 youths a year. These initiatives struggle to transform the lives of Nigerian youth because of their:

- Limited scope and scale
- Limited focus on the unique needs of young Nigerians
- Poor collaboration and cooperation across the sectors
- Outdated and theoretical training models
- Distance from the grassroots.

In the ' Next Generation Report ' of the British Council in 2010, it is highlighted that "Nigeria needs to develop the infrastructure that will underpin a world class economy, spending up to an additional 4% of Gross Domestic Product (GDP) on this task. It should diversify away from oil, with an emphasis on sectors that will improve employment prospects for young people, while removing obstacle s to economic growth and private enterprise. The oil industry contributes as much as 40% to national GDP, but is highly capital-intensive and employs only a tiny fraction of the population. Other industries still in their infancy offer greater potential to Nigeria and Nigerians: communications; manufacturing (textiles, clothing and footwear; automobiles); and the mining of resources other than oil." Clearly, national development cannot be spoken of without the citizenry first believing in a common goal that obliterates segregation between people of differentiated peculiarities in a community, this is critically supposed to be heaved on the responsibilities of the government. Poverty and lack, illiteracy and miseducation, bad leadership a nd poor governance are the threading of the society, that when weaved with the economy of the people, brings about dissociation from any that shares no particular quality with themselves.

However this has to be changed through changing the mindset of people through the quality of education received within the walls of our institutions. According to National Bureau of Statistics (2009:238; 2010:2), the national unemployment rates for Nigeria between 2000 and 2009 showed that the number of unemployed persons constituted 31.1% in 2000; 13.6% in 2001; 12.6% in 2002; 14.8% in 2003; 13.4% in 2004; 11.9% in 2005; 13.7% in 2006; 14.6% in 2007; 14.9% in 2008 and 19.7% in 2009.

Lack of entrepreneurial education, that creates a self-reliant ideology in people will lead to gross unemployment, which results in poverty and lack; this in turn creates survival instincts in individuals, and then brings about segregation among the people, Nigeria could never attain integration with the current educational strategy it breeds. Entrepreneurial education will however lead to increased employment, reduce poverty level, and bring about entrepreneurs working together for common good, therefore establishing a stronger economy than the present Nigerian economy. Entrepreneurs can take advantage of the nation's state to grow their businesses, create jo bs and employment for required skill sets they will find within their societies and grow. Poverty entails more than just the lack of income and productive resources to ensure sustainable livelihoods. Its manifestations include hunger and malnutrition, limited access to education and other basic services, social discrimination and exclusion as well as the lack of participation in decision-making. Various social groups bear disproportionate burden of poverty.

A social perspective on poverty should contribute to the debate on the effectiveness and limitations of current poverty reduction strategies, one of which is education. Education is critical to the development of personal economy from poverty level to at least the level of self-sustenance, however, the current curriculum of the Nigerian educational system, prepares the average graduate to be prepared to become employed by the available firms or business owners in the society, which in turn grows the number of un employed in the society. This emphasizes more on the quality of education administered, than the number of students that pass through school, according to the goals of the United Nations for developing countries. Among the barriers that would hinder progress within the current Nigerian education systems certainly are (these are however, not unique to the educational sector):

- Rigidity of systems: Certain people benefit from the status quo, and are resistant to change. Others are unable to see the possibilities for change, or lack the commitment, courage or energy needed.
- Governments, or local leaders, who are generally not held accountable for how much money is spent, and how education systems are managed.
- Sufficiency attitude – what is provided for the poor is good enough.

- Inadequate pro-poor infrastructure or support systems – this makes it difficult to implement successful poverty eradication interventions.
- A lack of systematic tracking of pro poor interventions – in this way it is extremely difficult to know if the activities and programmes implemented have had any impact at all.
- Nigeria cannot combat the ills of the society just by ra ising its budget; there should be a strategic systematic approach to education that exists, an d bridging the gap between its service delivery and its effectiveness in the country.

Education in Nigeria is devoid of the element crucial to averting the surging rate of unemployment in the country, therefore the breeding of psychological dependence on direct access to money. Entrepreneurial development through education will advance the economy of the nation; much credence should be given to it and ingrained with focus on profitable personal development. Unemployment prevails in the country, hence, the growth of violence, poverty and segregation amongst citizens, because the educational system itself fails to empower the ones passing it, therefore not catering to the economy its results should enhance by default. This should be the core message of the evolving educational policy of Nigeria which is devoid of a system of education that emphasizes on the need to culture the country's young through the knowledge of rudimentary entrepreneurial development, common cultural heritage, and identification of exploitable strengths of structures, systems and cultures of others. If anything, there is a greater need of its augmentation, which can propel the country to work towards enduring peace and harmony. The focus of this study brings to the fore the crucial need for entrepreneurial education in Nigeria, putting more consideration on the educational system, strategies and its eventual social developmental effect in the society. The study highlights the problems of the level of education i n the country and its equivalence to level of poverty in the society by virtue of lack of employment or knowledge of how to startup businesses; the resultant effect of which is a society that breeds healthy partnership and motivation, impacting on the development in the nation. One of the MDG goals highlights education as a critical factor to reduce poverty and dependency on in developed nations, however, the contribution of this thesis to knowledge identifies what is important to the economy, which is qualitative education focused on the needs of the economy per time, rather than the resolution of the United Nations to increase budgetary details, increasing the people that go through school. This is not the first paper on entrepreneurial education; however, it is the first to do a comparative study on what is being implemented by different universities to see the effectiveness and brings to light the results in order to ascertain which strategy would eventually work for the Nigerian economy in truly eradicating unemployment in Nigeria.

A consistent universal theory does not unilaterally exist for entrepreneurship, but it however consists of several different approaches including sociology, psychology, anthropology, regional science and economics. There is no common theoretical framework that exists to synthesize the different points of views. Some trials to develop multidimensional approach to entrepreneurship study the problems also mainly from perspective mentioned above in some well-established disciplines. In trying to combine economic and social context in the same multidimensional model we run into severe methodological difficulties. Many of the studies assume and generalize that all the entrepreneurs, including the owners of small businesses and high-growth innovative ventures, are similar. The diversity of entrepreneurial phenomena and new ventures should be taken into account in modeling entrepreneurship. However, no single taxonomy captures all of the important differentiating variables of entrepreneurship. But more accurate differentiation in categorizing entrepreneurs gives the possibility to combine different traditions and theories in approaching multidimensional phenomenon of entrepreneurship. Unemployment is one of the developmental problems that face every developing economy in the 21st century. International statistics portray that industrial and service workers living in developing regions account for about two-thirds of the unemployed. (Patterson *et al*., 2006).

The Nigerian economy since the attainment of political independence in 1960 has undergone fundamental structural changes and challenges. The structural shifts have however not resulted in any significant and sustainable economic growth and development. Available data show that the Nigerian economy grew relatively in the greater parts of the 1970s, with respect 16 to the oil boom of the 1970s; the outrageous profits from the oil boom encouraged wasteful expenditures in the public sector dislocation of the employment factor and also distorted the revenue bases for policy planning. This among many other crises resulted in the introduction of the Structural Adjustment Programme (SAP) in 1986 and the current economic reforms. The core objective of the economic structural reform is a total restructuring of the Nigerian economy in the face of population explosion (Douglason *et al*., 2006). Unemployment in Nigeria, in the history of Nigeria and the role education is supposed to play to reduce it, so that education does not produce graduates for exports, but for indigenous development in the country, reduce poverty, improve standard of living, improve productivity, and overall improvement in economic performance among other benefits.

Chapter 13

Internal and External Factors Affecting Small and Medium Enterprises

13.1 Internal and External Factors Affecting Small and Medium Enterprises

Margi Lévy and Philip Powell argued that growth is determined by a combination of the entrepreneur, strategy and the firm organization. The author also defined key influences on SMEs growth, internal factors that reveal how decisions and features effect the growth of a firm.

(a) Entrepreneurial influences

A firm's growth is dependent on the managerial knowledge (Macpher son & Holt 2007, 172-192). Training is crucial for the productivity and quality as well as it influences the effectiveness, efficiency and motivation of the employees (Thassanabanjong et al., 2009). The SME owner has considerable personal influence over a firm's strategies, tactics and operations to engage in decision process across the firm.

As a result, although a flat, informal organizational structure is likely to exist, decisions making tends to be quite centralized around the owner. The entrepreneurs' personality and behavior are to be causal factors for or against growth - orientated achievement. It is characteristic of small business that power decision are centralized at the level of owner - manager, so his or her personality, skills, responsibilities, attitude and behavior will have decisive influence on business strategy. (Levy and Powell, 2005).

Market competition has a larger affect for SMEs chance of survival and uncertainty is high as most of the smaller companies tend to have smaller share of the capital. This leads SMEs to have one or two major customers and are do not have big influence on prices. Larger companies with higher market share usually determine prices. (Levy & Powell, 2005).

(b) Education of employees

SMEs face difficulties employing and retaining skilled graduates, because they prefer to work for LEs that can offer higher salary, job security and career possibilities. In order to meet the demands of the fast changing work environment which is typically associated with SMEs it is essential that smaller firms ensure that they are able to attract, retain and motivate high quality employees with effective transferable skills through the existence of a strategic training plan and a specific budget for training (Jameson, 2000).

(c) Marketing

According to Brush et al. (2009) marketing is another obstacle for companies to grow since many businesses confront challenges establishing effective distribution channels, communicating product features, pricing products and services in an attractive way, implementing sales and marketing efforts to win and retain customers and undertaking constant product development in order to sustain sales. SMEs generally do not have the knowledge or information about other markets, thus, this limit their ability to market their products to larger groups of customers and expand their business.

(d) Technology

SMEs tend to have low productivity and they are weak in terms of competition which is the result of using inefficient technology, not maximizing machinery utility and not improving in technology due to the limitation of funding and most SMEs are mainly users of technology, not adaptors of technology (OSMEP, 2007)

The World Bank (2009) claims that investments in technology are required in order to build up existing capacity and to improve the quality and productivity of production which will generate in higher value - added products that will improve the competitiveness for firms.

(e) Innovation

Innovation also plays a crucial role in now day business and it is regarded as a key characteristic of SMEs, mainly due to the attitude of the manager. Innovative companies are able to respond within the bounds of the knowledge about existing products or services to changes required by the customer within their niche market. (Levy & Powell, 2005). The impacts of globalization have pressured SMEs to greater demands. Innovation relies on bringing together different types of research and utilizing this knowledge to design new products, therefore innovation greatly relies on research and development investments. Normally developed countries allocate about 3% of GDP to R&D activities. Some developing countries, including China, India and Brazil, have rapidly increased their R&D expenditure, to levels with those of the world's most developed countries (Morrison, 2006).

(f) Financial support

According to Guffey business plan is essential when you start your own business. Unless you can count on the bank of your relatives you will need financial backing such as a bank loan or venture capital supplied by investors. A business plan is critical for securing financial support (Guffey, 2008). Throughout the region, SMEs do not have an easy access to credit and equity finance. This is because of the weak banking institutions in the region, the absence of capital markets, and the weak legal framework for credit and collateral. Finance in general are critical issues for growing businesses, forming the primary resource base from which other factor inputs are acquired. There are various ways the business owners can finance the growth of their firms but the fundamental decision is whether or not to accept external equity finance return for part ownership of the business. If owners allow external equity finance they choose to relinquish part of their control to either a financial institution or other individuals. Financing the firm is essential and getting access to finance plays a crucial role on firm's growth process. For many lenders it is almost impossible to assess the risks of an investment this is mainly because of the high level of uncertainty. (World Bank, 2001).

13.2. Agriculture Growth Factors

Agriculture is ranked as one of the extraordinary success stories past decades. In poorer countries with lower productivity rates and growing populations, agriculture continues to expand into marginal and fragile lands. Helping devel oping countries to intensify their agriculture may seem the most obvious and sensible solution but intensification at least under current models carries significant stings in its tails (Steiner, 2006). The lacking point here is that 60 percent of ecosys tem services - such as fresh water, air and water regulation, and the regulation of region climate are being degraded or used unsustainably (Steiner, 2006). Agricultural development plays a crucial role to overall economic growth in many developing countries. As farmer's incomes rise, so does their demand both for farm inputs and services, and for non-farm goods. There are more facilities processed if agricultural production increases (Båge, 2006).

The growth and development of agriculture requires investment and technology. Through advanced technology huge productivity gains are possible. Despite sustainable improvement over the years, the transfer of technological achievements in the agricultural sector, remains as low as around 40 per cent. Comparing this with developing economies it is estimated to be only half of that. The low level of technology transfers has not only wasted valuable resources of technology available to the agricultural sector but has also hindered the development of agricultural and urban economy (Xuedong, 2006). The main technological challenge is to develop in a way that is relevant to small farmers and to create the conditions they need to transform their small plots into viable small business that makes a vital contribution to local and national economies (Båge, 2006).

Increased spending on agricultural research is vital, but it is equally important to ensure that the research carried out benefits the smallest farmers. The competition is intense in agriculture business. Water is an issue, because it is used in many different purposes. Water is used for economical purposes such as growing food and energy production. Poor people needs for drinking water to compete with these economic uses in developing countries. When water competition gets heavy usually the suffering part is environment (Catley - Carlson, 2006). Agricultural policy has often developed without concern for principles and competition policy. Historically agricultural policies have been primarily devoted to improving the welfare and agricultural procedures, because of the political imperatives and social values (OECD, 2004).

13.3. Construction Growth Factors

The construction business most generally starts by ambitious construction project managers and executives. Managing construction business differs from the others and most of the times managers are not likely prepared to take on the range of responsibilities forced on them in managing the business of construction in its entirety. (Ganaway, 2006) Ganaway (2006), in his book Construction business management: a guide to contracting for business success clarifies requirements for managers before establishing construction company. The initiatives are import ant to know you are capable to manage yourself the company. Passion plays crucial role in field of construction. The construction industry is seen as a highly turbulent an, rapidly changing, very complex, and extremely competitive. (Chen & Mohamed, 2007).

According to Chinowski and Meredith there are numerous factors that have impacted construction industry in the past decade. The first is the emergence of knowledge-based rather than skill based environment. The focus is on expertise in areas such as technology, economics and market development. In the past the focus was more to remain with a single employer for an extended period of time in the tradition of professional apprenticeships. The second factor influencing in companies growth in construction industry is the information revolution e.g. new technique, systems and platform. The third essential factor is the rapid change of construction industry and competition. The tough competition has forced several companies to improve efficiency and cost effectiveness. (Chinowski and Meredith 2000). According to Langford and Male companies' growth can be achieved in many ways. Through growth in size where managers' focus is on the external environment in order to pursue opportunities and less on internal efficiencies or growth in size and efficiency where the attention is on external opportunities that are presented in the business environment and internally to increase efficiency.

Chapter 14

Comparison with Social Entrepreneurs to Corporate Entrepreneurs

14.1. Characteristics of Social Entrepreneurs Compared to Corporate Entrepreneurs

Prior research has found many parallels between social and commercial entrepreneurs when it comes to personality and behavioural traits (Drucker, 1989; Gitman & McDaniel, 2008; Gordon, 2006). There are some essential personal characteristics of these entrepreneurs that contribute to the success of the venture. A recent empirical study conducted by Sharir and Lerner (2006) identifies eight variables that contribute to success of a social venture. Out of these eight variables, five are directly related to the qualities of the entrepreneur, namely, the entrepreneur's social network , commitment , previous management experience, ability to integrate the vision and to establish strategic alliances . Similarly, a study of the characteristics of social entrepreneurs in the United States found that social entrepreneurs are more likely to have high social capital (Ryzin, Grossman, DiPadova-Stocks & Bergrud, 2009). Although such studies of the personal characteristics of social entrepreneurs provide useful information about the leadership and organisational skills of such individuals, they do not explain the essence of these traits, why they are important and how social entrepreneurs use them to achieve their social mission and create and sustain social change. More research is required to understand what values, motives and behavioural repertoires distinguish a social entrepreneur from a corporate entrepreneur.

Several other studies of social entrepreneurial traits and qualities have explored how such personal, social, and organisational skills contribute to the success of the SEV. For example, one key attribute of social entrepreneurs identified in prior research on the topic is entrepreneurial credibility. Reputation or credibility is vital for social entrepreneurs to be able to tap into their social networks to garner and mobilise resources. Credibility does not necessarily mean personal charisma, but rather the ability to assemble and effectively utilise many resources (Waddock & Post, 1991). The ability to develop a network of relationships is a hallmark of visionary social entrepreneurs (Thompson *et al.*, 2000). Kramer (2005) argues that people identified as social entrepreneurs are often under scrutiny for their capabilities, character, and leadership abilities by their target population. Personal credibility is equally important for comparison although their professional credibility is what matters most as their business ventures evolve over time. Credibility is a key factor for social entrepreneurs as it also helps to maximize others' commitment to the collective purpose championed by the leader. It is precisely the leader's credibility that helps social entrepreneurs to maintain a clear focus on the overarching goals (or vision) of the organisation (Waddock & Post, 1991).

Orloff (2002) identifies 'high quality leadership' as an essential prerequisite for the emergence of a social venture partnership and its continued success. Transformational leaders 9 motivate people to achieve transcendent or end values such as liberty, social justice, and equality (the ends over means), whereas transactional leaders 10 motivate people or followers to achieve modal values such as honesty, responsibility, fairness and honouring commitments (the means over ends) (Burns, 1978). Similarly, Waddock and Post (1991) pos it that in order to achieve the value embedded in the collective vision it is necessary for a social entrepreneur to embrace end values rather than modal values. Such a leadership style can attract followers who take up the SEV's mission and social values and carry forward or enhance them realising the common vision of the SEV and fostering collective purpose. 11 Commitment to collective action is essential for holding the organisation together and enhancing feelings of community and valueadded collaboration among employees and/or volunteers. Decision-making power is more distributed as transformational leaders rely on collective wisdom, experience of the community, employees as well as partners, and understand the importance of collaborative capacity building. Corporate entrepreneurs also have to demonstrate a similar type of leadership style. However, while intrinsic motivation among followers or collaborators is necessary for a social entrepreneur to maximise the benefits of the venture, extrinsic motivation may be sufficient for their counterparts. Efficiency and effectiveness seem to be more essential for corporate entrepreneurs, while value-added collaboration is more important for social entrepreneurs. Decision-making power is also rather limited to the management in corporate ventures. All of these personality and behavioural characteristics should be systematically and empirically assessed in future research on the distinctive features of social and corporate entrepreneurship.

14.2. Distinguishing Processes of Social Entrepreneurship as Compared with Corporate Entrepreneurship

Socio-environmental problems are inherently complex and social entre- preneurs are very adept at recognising these complexities. Waddock and Post (1991) argue that social entrepreneurs not only have a unique ability to recognise the complexities of socio-environmental problems but they are also able to frame the problem in a new way that increases public awareness of the problem through their vision. They identify situational multiplexity, crisis (relevancy of the problem) and interdependence (requirement of multiple collaborators) as factors that lead to innovative vision. Whereas innovation, competition and profits are the driving forces for commercial entrepreneurs, social entrepreneurs foster innovation and inclusiveness, which enable them to bring about a positive change in the system and the society (Jeffs, 2006).

Social entrepreneurs are usually supported by volunteers who share a common vision of mitigating socio-environmental problems. Since profit generation or maximisation is not at the centre of such ventures, resource mobilisation for social enterprise is very different compared with tradiitional businesses? Social entrepreneurs are the main propellers of resource generation in their organisation. Along with their personal wealth, they rely heavily on their social networks to carry forward their mission. Social entrepreneurs understand the necessity of being inclusive to generate a feeling of ownership and sense of value-added participation 13 among collaborators. Value-added participation is at the heart of collective social action (Waddock & Post, 1991). Peredo and Chrisman (2006) also emphasise the importance of inclusiveness. They argue that lack of ownership is the main reason that many poverty alleviation programmes have devolved into global charity since most projects are conceived and managed by development agencies rather than members of community.

A different type of inclusiveness is however encouraged in commercial enterprise. Since the product/service offered by commercial entrepreneur is either new or more technologically advanced than other similar products/ services that are available in the market, the ownership (trade secrets or intellectual property) are generally closely guarded. However, inclusiveness is encouraged in the form of vertical/horizontal integration through the supply chain management process, where everyone involved in the production of the product or service, from the raw material manufacturer to the distributors of final product, are all working in coordination to achieve economic efficiency and effectiveness. Some corporate leaders are able to adopt a hybrid style of leadership whereby they are successful in creating a sense of psychological ownership for the organisation among its employees.Van Dyne and Pierce

(2004) highlight many positive influences of sense of ownership on employee morale, motivation, behaviour and productivity. However, it is important to note that this leadership style is pivotal for a social entrepreneur to ensure the success of the venture, but not for a corporate entrepreneur.

Commercial entrepreneurs are relatively less bounded by resource constraints. They have varied sources of financial support available to them based on their economic proposition (e.g. the market, venture capitalists, banks), and they can afford the best human resources available. Personal social networks are equally important in the initial stages of a new business venture, but the focus shifts to professional networks as the venture grows.Another point of contrast is the issue of long term financial planning in both kinds of ventures. Whereas strategic financial planning is often overlooked by social enterprises, it is of vital importance in commercial ventures. Yet, the importance of strategic financial planning is gaining importance in the social entrepreneurial world to ensure a revenue stream so essential for sustaining the venture (Boland, 2002; Boschee, 1997). In SEVs, the social entrepreneur's sustained efforts are required to garner funding for his/her social cause unless she is able to create self-sustaining financial system whereby others (including community members) can take charge of the management of the initiative and generate sufficient resources to sustain it.

Many scholars have written about the importance of business planning (Massarsky & Beinhacher, 2002; Rouson, 2005; Zietlow, 2001). They argue that sound business planning has a significant impact on the success of a venture. Business planning assistance, in the form of targeted business analysis, market research and strategic planning, could be a valuable resource for non-profits and social enterprises. The question, however, is how can 'business planning' capacities be provided to social entrepreneurial organisations since these organisations vary widely in terms of their missions. Further, Dees (1998) and Foster & Bradach (2005) caution about the erosion of social concerns with the adoption commercial approaches to self-sufficiency as they might create unrealistic expectations from entrepreneurs and distort managerial decisions, waste resources, and leave important social needs unmet. Boschee (2008) supports this argument as he explains that lack of skills, resources, and the proper mindset can compromise the social mission of the organisation. These scholars (i.e. Dees, Foster & Bradach and Boschee) also stress the need to explore all strategic options 15 including their ability to use the social mission to tap into the financial resources of philanthropic foundations and charitable organisations. Corporate enterprises are well positioned when it comes to financial sustainability as they typically rely on business development techniques and strategic planning. Recently venture philanthropy 16 has gained importance, reflecting a paradigm shift away from the notion of a social sector that merely receives funds from charitable organisations, towards a notion of earned investment through collaborative partnerships (Boland, 2002).

Similarly, the sociocultural context of a corporate or social venture is very important in influencing the success of both types of entrepreneurs, despite the differences in their goals. The success of an SEV largely depends on the local political, social and cultural context, all of which influence how the local community perceives a social problem and its proposed solution. Similarly, histories of success of similar ventures as well as the credibility of the entrepreneur also influence the outcome of the SEV (Waddock & Post, 1991). Social entrepreneurs understand the situational multiplexity, the relevance of the social problem, as well as the unique contextual circumstances surrounding and influencing the social problem and define it appropriately to generate awareness of the problem among others by promulgating an innovative vision. Social problems exist primarily in terms of how they are defined and conceived within a particular society. Blumer (1971) argues that the societal definition of social problems determines their life cycle, how they are approached, and what is done about them. He adds that under- standing the processes by which a society comes to see, define and handle a social problem is extremely important since the social problem is always the focal point for the operation of divergent and conflicting interests, intentions and objectives. This interplay of interests determines the ways in which society deals with the problem. Thus, a broad understanding of influential contextual factors is a key to the success of SEVs. Commercial entrepreneurs, by contrast, are concerned with broader macro-level contextual factors such as the macro-economic environment, tax and regulatory structure, technological advancements and the larger socio-political environment related to labour, religion and politics (Peredo & Chrisman, 2006).

14.3. Key Outcomes of Social Entrepreneurial Ventures Compared with Corporate Entrepreneurial Ventures (Cevs)

Social entrepreneurs generally try to strike a balance between social and economic value creation through an innovative, visionary approach towards addressing the social problem. They differ from NPOs in that they aim to reverse an imbalance in society or the community by creating sustainable positive social change. NPOs, on the other hand, may or may not aim for sustainable positive social change. In contrast, corporations can have several alternative secondary goals such as improved customer service, corporate social responsibility and improved brand image and value. Because social enterprises tend to be member-controlled and people- centred, and the benefits of their activities are frequently non-monetary in nature, they aim to contribute to the maintenance of economic and social cohesion in the community (Oatley, 1999). Since such benefits are challenging to measure, it is difficult to gauge the impact or effectiveness of social enterprises as compared with commercial enterprises, in which the expansion of personal and stakeholder wealth is the ultimate goal.

Apart from the benefits available to the target population (section of the population that the SEV directly serves), both types of entrepreneurial ventures can benefit the larger community. Commercial ventures can enhance the prosperity of the community by providing job opportunities and infrastructure (e.g. roads, highways, water and waste management) to community members. SEVs, on the other hand address deep-rooted social problems, social injustice or societal imbalance through processes of social change creation and sustenance. Social enterprises mobilise interpersonal and professional networks to create economic and social capital by encouraging citizen engagement and empowering individuals and communities. The difference, however, is that a commercial enterprise will only continue its operation to the extent that the venture is economically viable, whereas SEVs are likely to continue their activities whether they are economically viable or not as they do not aim to increase personal and stakeholder wealth. Both kinds of ventures create different forms of social value. For commercial enterprises, however, social value creation is not the primary motive, whereas for SEVs it is the primary reason for their existence. Furthermore, commercial ventures are limited in their capacity to create social value. Commercial enterprises generate social value indirectly by generating economic gains, often by bringing resources (e.g. materials, manpower, human capital) into the organisation and creating financially valuable outputs (e.g. innovation in the form a commercial product or technology) for sale outside the organisation.

Therefore, the social value created by commercial ventures is intrinsically tied to economic value generation. SEVs, on the other hand, are committed to creating social value within and outside the organisational boundary. For example, social enterprises encourage the sharing of ideas, innovations and best practices with other social enterprises, non-profits, NGOs and some commercial enterprises. They encourage collaboration as opposed to competition with other organisations and foster the creation of knowledge and social networks. In this way, social enterprises spread their resources outside the organisational boundary. Moreover, social enterprises create outputs that may or may not be linked to economic benefits for the organisation (e.g. One World Health produces vaccine for black fever that saves many lives in developing countries even though the market value for this vaccine is low).

Social value is a concept that is difficult to define and measure. Understandably, the evaluation of the processes and outcomes of social entrepreneurial efforts is a challenging task. Some of the most discernible problems in this regard are:

(i) how to measure social outcomes? (ii) is it even possible? (iii) what timeline should be used to evaluate such efforts? (iv) how to measure indirect benefits? and (v) is it morally justifiable to evaluate benefits that might not be measurable in the first place through a cost-benefit analysis and attach a monetary value to them? In such a scenario, it is pertinent to ask whether it is fair to compare social entrepreneurial efforts with for-profit entrepreneurial efforts. It has been argued that the social marketplace (i.e. the social sector consisting of non-profits and civil society organisations) is generally less demanding of entrepreneurs in the sense that it neither rewards superior performance nor punishes inferior performance as readily as the commercial marketplace (Austin *et al*., 2006). Yet, such comparisons may be meaningless given the fundamentally different motives of these organisations. The fact that social and corporate entrepreneurs differ on what they perceive as an 'opportunity' suggests that little common ground exists for such comparisons.

Moreover, attempting to measure social impacts or outcomes through the cost-benefit lens amounts to imposing an economic value on some of the most unquantifiable outcomes. It is unfortunate that the major criterion typically invoked when evaluating entrepreneurial ventures is the economic one. This is not to suggest that appropriate measures of effectiveness should not be developed, but the need for new measures is dire when funding decisions are guided by such incomplete and inaccurate measures. In a recent report from the Bill & Melinda Gates Foundation's Impact Planning and Improvement Division, Tuan (2008) identifies important limitations associated with various cost-benefit criteria for estimating social value. Some of the major concerns of using such methods include: (i) use of a monodimensional yard stick for measuring inherently dissimilar outcomes, (ii) the necessity of 'informed guesswork' to yield cost-benefit estimates, (iii) unintended consequences of using measurable proxy variables for unquantifiable constructs, (iv) the costs of measurement, and (v) the inability of corporate cost-benefit measures to capture value creation outside the boundaries of the organisation. The dangers of using such evaluation methods are the lure of false precision, poor construct validity and focusing on a single numerical value or index to indicate whether or not an investment has been successful.

Milton Friedman, among other neo-classical economists, asserted that the only social responsibility of a business is 'to use its resources and engage in activities designed to increase its profits through open and free competition without any deception.' (Friedman, 1970).There is numerous criticisms of such pro-capitalist views. For example, Rossouw and Van Vuuren (2004) argue that by its very nature a business needs to serve the interests of society. According to them, businesses derive their ultimate justification not from economic objectives, but from the moral objectives they pursue. Drucker (1979) argued that even the most private penterprises is an organ of society and serves a social function. Moreover, increased stakeholder activism has created an environment where businesses can no longer afford to focus solely on profits and are forced to be more responsive to societal needs. A new area of social impact management has emerged that focuses on the intersection of business practice with wider societal concerns that reflects and respects the complex interdependency between these two realities. Gentile (2002) argues that this is a critical part of the contemporary business because without understanding this interdependency, neither businesses nor the society in which they operate can thrive.

The idea of balancing social value with economic value has created an impetus among businesses for jumping on the social enterprise band- wagon. Corporate ventures have been quick to respond to the concept of social entrepreneurship with the idea of Corporate Social Responsibility (CSR) to cast off some of its burdens of social costs while achieving what is known as the 'double bottom line' (Doane, 2005; Pendleton, 2004). The result has been the emergence of hybrid organisations classified as 'socially responsible businesses' defined as, 'a venture (generally for-profit) that seeks to leverage business for a more just and sustainable world' (Social Network Venture).

Every year, Corporate Responsibility Officer (CRO), a business ethics magazine, produces a list of the 100 best corporate citizens (i.e. companies). CRO's criteria for exemplary corporate citizenship include positive action in area of environment, climate change, human rights, employee relations, philanthropy, finance and governance. Some of the most infamous companies have made it to this list such as Exxon Mobil, Monsanto (guilty of many unethical practices, including child labour in India), Gap (responsible for employing child labour in India), Citigroup, and Goldman Sachs Group, Inc. (blamed for outrageously high CEO and executive compensation), Coca-Cola (accountable for ground-water depletion and pollution in developing countries), and Dow (responsible for the Union Carbide gas leakage in Bhopal, India; (http://www.global exchange.org/, www.bhopal.com). It is, therefore, pertinent to question the validity of CRO's measures of CSR.

Doane (2005) argues that CSR is only a tool to reap the financial benefits of ethical consumerism for certain corporations. She explains that the problem with the assumption of doing well while doing good (also called the 'double bottom line) is that markets do not really work that way, as any investments towards a social cause are not considered 'wise' because they stand contrary to the profit-making motive fostered in capitalist societies. Similarly, Bakan (2005) asserts that just as human psychopaths disguise their dangerously self-obsessed personality by their ability to use charm, CSR may play the same role for corporations. CSR fails to recognise that it is the institution of the corporation itself that may be at the heart of the problem. According to Doane, the CSR movement is winning the public relations game with both the govern ment and the public by lulling us into a false sense of security. In the end, CSR is nothing more than a placebo, leaving us with the immense and ever-increasing challenges of globalisation and corporate capitalism for the foreseeable future. Neo-classical economist and a Nobel Laureate, Milton Friedman, believes that there is only one instance when CSR can be justified when it pursued by a corporation — when it is insincere. He adds that such a strategic view of CSR reduces lofty ideals to a hypocritical window dressing, but it is virtuous when it serves the bottom line (Bakan, 2005).

Contrary to Doane and Bakan's valid criticisms of CSR, some businesses are known to follow ethical and environmentally sustainable business practices without any hidden agenda. Examples of some such businesses include Green Mountain Coffee Roasters, Nordic Naturals (Alter, 2004) and Dean's Beans because of their environmentally sustainable and fair trade practices. It is, however, necessary to clarify that such organisations cannot be classified as SEVs even though they may share some common features, since the primary motive of a business enterprise, albeit a socially responsible one is profit maximisation and not the mitigation of a social problem. For SEVs, on the other hand, financial profit serves as a means towards achieving the primary goal rather than an end in itself (Schuyler, 1998). Similarly, for-profit firms that are in the business of producing and selling high social-value creating goods (e.g. Pharmaceutical companies that produce and sell drugs) cannot be con- sidered socially responsible organisations as the decisions of such organisations are strictly guided by the profit motive. We here summarises the earlier-described distinguishing features (i.e. motive, primary and secondary goals, antecedent conditions, outcomes, role of the social entrepreneur, collaboration needs and value creation) of social enterprises, corporate enterprises as well as NPOs and NGOs.

Social and corporate enterprises were compared on the following dimensions: (i) the purposes for their existence, (ii) the role of the entrepreneur during the lifecycle of the venture, (iii) the entrepreneur's personality and leadership traits, and (iv) the essential outcomes of the venture. Key conceptual differences between social and corporate entrepreneurs and the practical implications were identified and discussed. In doing so, new questions were raised and hypotheses were posed for empirical investigation in future research on this topic. For example, it is hypothesised that a transformational leadership style is essential for the success of a social entrepreneurial venture but not for a corporate venture. Additional priorities for future research identified in this article are to:

(i) develop new constructs and measurement tools for evaluating the processes and outcomes of SEVs; (ii) explicate the influence of antecedent conditions in the creation of an SEV; (iii) determine how contextual factors facilitate or hinder the success of an SEV and the relative importance of these contextual factors; (iv) explain how SEVs leverage their social mission to ensure financial sustainability; (v) document the value of social and professional networks for achieving the SEV's social mission; (vi) examine the influence of an entrepreneur's personal and pro- fessional credibility on the success of an SEV; (vii) analyse how various personal, social, and organisational skills contribute to the success of an SEV; (viii) understand how SEVs foster innovation and inclusiveness and create social value within and outside their organisational boundaries; and (ix) conceptualise and elaborate the construct of social value or social impact as a basis for developing new methods to evaluate it. A broader and more fundamental issue raised in the article is the necessity of separating the scientific study and evaluation of social entrepreneurship from management and economic philosophy in order to develop a more nuanced and accurate conceptualisation of social entrepreneurship and social entrepreneurial ventures.

14.4. Relevance of IB Models For (Private Sector) Development

Inclusive business models have the potential to create win-win solutions for profit-driven as well as social interests. Often, governments do not have the resources, the capacity or the will to provide necessary goods and services in poor or remote areas. In contrast, IB models can offer the capacity for innovation which is needed in order to serve areas that are remote and are traditionally deemed inaccessible. They have potential to provide access to products or services, and to increase the purchasing power of the poor by offering relevant goods and services at lower prices, or with the use of payment schemes that improve affordability. They can also create employment and income by integrating the poor in the value chain as distributors, suppliers or employees. Moreover, the underlying profit-orientation helps to ensure the long-term financial sustainability of initiatives beyond the duration of donor-funded projects. It is for this reason that development agencies working in energy, water or health – to name just a few sectors – are promoting the increased use of partnerships with the private sector or with appropriate NGOs, supporting them in establishing IB models. At the same time, especially when they focus on IB models in local companies and the integration of local producers and micro-enterprises, IB promotion measures can provide additional benefits in terms of pro-poor private sector development (PSD). The objective of PSD is to strengthen the private sector – mainly micro, small and medium-sized enterprises (MSMEs) – in partner countries, in order to create employment, increase incomes and raise government revenues through tax payments, all of which will contribute to poverty reduction and welfare creation.

Promoting IB models in partner countries can contribute to these objectives in the following ways.

- **Increased employment and income opportunities through start-ups:** A key constraint for MSMEs in developing countries lies in the strong competition in markets for simple consumer products and services. Local companies offering new products and services intended to meet the needs of poor people can tap new markets and thereby generate higher revenues, while increasing incomes and employment for themselves.
- **MSME upgrading through better services and inputs:** IB models can increase the productivity of the poor by providing infrastructure (e.g. electricity, health services), knowledge and market access (e.g. to finance, agricultural inputs etc.) and by reducing transaction costs.
- **MSME upgrading through integration in (regional or international) value chains:** IB models can improve market access for small producers and micro-enterprises by integrating them in national and international value chains (supply-side IB models). This enables them to attain increased and more regular income. Delivering information, training and other services, such as financial services, to small-scale suppliers offers them an opportunity to upgrade.
- **More start-ups and business growth by improving the business environment:** IB models that can prove the market viability of supplying basic services to poor people could, if accompanied by dialogue with the public sector, stimulate sector-specific reforms. This in turn would promote a generally more open and competitive business environment in which the private sector is able to provide basic services.

Chapter 15

General Barriers

15.1. General Barriers and IB Strategies to Overcome

There are extensive barriers when it comes to doing business with people at the base of the pyramid. This is because of the market inefficiencies and high transaction costs that can render businesses unprofitable, particularly in the short run. For instance, customers in remote rural areas are hard to reach as the necessary infrastructure is not available. General barriers to IB models include:

- **Shortages of adequate infrastructure** such as roads, energy, water etc., particularly in rural areas and urban slums.
- **The need to create a market (i.e. encourage demand):** many goods and services are known as 'push' products (e.g. waste management and preventive healthcare). It is necessary to raise awareness about them in order to stimulate demand; new distribution channels must also be created.
- **The lack of knowledge and skills** among the poor, to act either as clients or as suppliers and employees.
- **Complex or hostile regulatory or legal environments,** affected by weak institutions and insufficient enforcement of rules and contracts. IB models overcome these barriers through innovation and by employing additional, targeted measures. Indeed, the strategies they use to overcome barriers to doing business with the BoP are one of the main defining features of IB models. These measures can be categorised as follows.
- **Adaptation of products and processes:** For example, health and banking services are provided via the mobile phone, and sector-specific technical solutions have been developed to provide off-grid energy production, water purification or sanitation services.

- **Reinventing business processes helps** to circumvent barriers to IB, for example by adjusting to low-level cash flows by offering small-unit pricing or pay-peruse schemes, by providing shared access, or by selling through informal retailers. Through contract farming and 'deep procurement' that bypasses intermediaries small-scale farmers are integrated into value chains.

- **The benefits to the potential BoP business partners** must be clearly communicated, including indications, for instance, of what unmet needs will be addressed or how their quality of life will improve. Thus they can see that it is worth their while diverting the limited capital they have available. Effective communications strategies take the local context into consideration, with its specific challenges such as illiteracy.

- **Investment to remove market constraints:** IB models include investments in market research, education for the workforce, training of suppliers, building private infrastructure (grids, pipelines), provided those additional investments promise to bring benefits that are tangible and can be internalised for a period time at least. For instance, investing in small-scale suppliers by providing them with equipment or training will enable them to deliver reliably, and in the required quality and quantity. To achieve this, IB models may also include the leverage of additional funds from non-private sector stakeholders, such as the government, NGOs or donor agencies.

- **Integration of the poor in product development and delivery:** Integrating local communities in the work of awareness creation, education and service delivery helps to reduce costs in the long run and improves outreach to the target group as it increases trust and inclusion in local networks. Integrating the target group in the design and development (co-creation) of the products and services also helps adapt them more directly to people's needs, thereby increasing demand levels and the willingness to pay.

- **Engagement with the government in policy dialogue:** As IB models are often applied in highly regulated markets (health, education), or are used to develop new markets (micro-finance), they are likely to face regulatory constraints. For instance, if an inclusive business model aims to introduce a new innovative product or service to a highly regulated market such as finance or insurance it can face substantial regulatory challenges. In order to improve the regulatory environment in their very specific sectors, companies frequently engage in a policy dialogue with the government. This can be done individually, provided a company has sufficient visibility and market power, or through collective action with other stakeholders.

- **Expansion of the target group:** Instead of focusing only on the poor, adjoining income groups can be included in order to reduce the risk .

15.2. Organisation-Specific Strengths and Weaknesses

In addition to the general barriers to doing business with the poor, organisation-specific strengths and challenges need to be considered. Various types of organisation apply IB models, including companies and NGOs. As this book is about private sector development, we will only look at the strengths and weaknesses of different types of company. These can be categorised as 'local' or 'foreign', 'small' or 'large'. All types of company have their different strengths and weaknesses with regard to applying IB models. Due to their strengths and weaknesses, companies – local or foreign, small or big – all differ in their ability to apply the various strategies to overcome the barriers described above:

- Local SMEs have an advantage when it comes to understanding the needs of low-income clients, and they also benefit from being well anchored in the local community. Their core business often involves selling to, and sourcing from low-income populations. This means they are adept at developing low-cost products and services as well as adequate distribution strategies.

- On the other hand, SMEs often have little access to adequate human, information and financial resources, and lack strategic planning capacities. Individuals and SMEs with innovative ideas rarely possess the right skills to develop a sophisticated business plan, and it may be hard to attract financing in the early stages due to the high risk of new, unproven IB ventures. Related problems occur when the amount of capital needed falls below the investor's threshold, or when there company lacks collateral. At present, therefore, from the point of view of investors there seems to be a shortage of feasible, replicable IB opportunities. This is corroborated by the fact that in some countries there is an abundance of capital ready to finance IB ventures, although there is a dearth of applications based on proven, sufficiently large and financially sound business models.

- Moreover, having little access to technologies and information beyond the local context can hinder a company's capacity for innovation. Operating with a low profit margin can slow down growth processes and make it difficult to scale up operations.

- Large national companies or emerging market multinationals seem quite well positioned to invest in IB models. They combine a good general understanding of clients' needs, established local supplier networks, widely known brands, and adequate human and financial resources. They are used to conducting business in a difficult environment, and national companies in particular are well-connected locally.
- However, such companies might still support in increasing the social value and the 'inclusiveness' of their products and services, in finding civil society partners, in further adapting their products and services to customers' needs, and in unlocking additional financing or engaging in the policy dialogue. As the potential partners and suppliers for doing business with the BoP tend to vary from country to country, large national companies or emerging market multinationals will need support to establish strong networks when entering new markets.
- Multinational companies from industrialized countries are at the other end of the spectrum. They tend to be well equipped with human and financial resources, as well as R&D capacities. This means they are able to make more substantial investments to overcome barriers, adapt products and services, and engage in policy dialogue.
- However, they are often unfamiliar with the ideas and approaches of IB models, and they might question the effectiveness of such models for them, particularly when they are pursuing short-term objectives to maximise shareholder value. The concept of inclusive business often runs counter to established internal processes and standards, such as minimum profit margins or maximum investment horizons. Other challenges include the lack of market information, and poor understanding of the peculiarities of the low-income target market (such as consumption patterns, informal competition, weak supply structures). These larger multinationals also encounter practical challenges in implementation, such as language barriers or cultural differences. Private sector development (PSD) programmes wishing to focus on the wider use of IB models in order to achieve (private sector) development objectives can work with any of these types of company. They all have the potential to improve living conditions for the poor and to increase productive capacities, incomes and employment in developing countries.
- High-level conferences and forums bring together decision makers from private industry, the public sector and the donor community. Examples include the CEO forums hosted by SNV Netherlands Development Organisation 17, and the WBCSD (World Business Council for Sustainable

Development)18, which aim to promote IB models by prompting interest and commitments on the part of the participants.

- Awards: A number of competitions are used to identify and promote innovative IB models which have potential for scaling up and replication. They stimulate initiatives, generate and present best practice cases for replication, create awareness and select examples with high potential for further support. One example was the G20's 'Challenge on Inclusive Business Innovation'19, launched during the G20 Summit in 2011. Entrants had to prove the innovation of their business models, show the development results and the potential for growth and replication, and also demonstrate the social, environmental and financial sustainability. Case writing competitions such as those organised annually by EFMD (European Foundation for Management Development) 20 and sponsored by the UNDP's GIM Initiative (Growing Inclusive Markets) 21 serve to identify and share successful business models.

- Case studies are used to present examples of successful IB models. Examples include the reports from UNDP and IFC, and the case studies of the World Business Council for Sustainable Development (WBCSD), which even provides a case study template.

- Sector-specific practitioner guides create awareness and provide guidance to companies on developing and implementing IB models. They also include examples and lessons learnt.

- Websites and blogs: Blogs are a good source of information, opinions, case studies and publications. They include www.inclusivebusiness.org, which is maintained by SNV Netherlands Development Organisation, and the blog of WBCSD.

- Facilitating and moderating partnerships between companies interested in inclusive business, and potential partners. Apart from the activities described above, which not only create awareness but also facilitate contacts, donors organise special events to link individual companies interested in IB with partners that can provide support. These include governments (to create an enabling environment), investment funds (provision of finance), NGOs (as local implementation partners) and international donors (support in the form of training, services, grants). One example is 'Business Call to Action', an alliance of bilateral donors, multilateral organisations and the private sector, which aims to create awareness and linkages. Initiatives also exist to help companies of the global North to identify IB opportunities and local partners in developing countries. This approach is supported, for instance, by the Danish development agency DANIDA.

Chapter 16

Medium and Small Enterprises – An Overview

16.1. Overview of Medium and Small Enterprises (MSE) Growth

In most developing countries, MSEs constitute the vast majority of firms, generating a substantial share of both overall employment and output. Given their significant economic role, one might expect MSE growth to drive overall increases in output and income levels. In many cases, however, their largest economic contribution appears to be one of maintaining, rather than generating new, employment and income for the poor. At an aggregate level, MSEs demonstrate impressive growth, 3 especially when compared with larger firms. However, many individual MSEs grow slowly or not at all, in some cases, due to a conscious decision on the part of the business owner. Overall growth rates are often fueled by the rapid expansion of a narrow group of highly performing MSEs. In dev e loping as well as in developed countries, there appears to be a small group of "gazelles", firms that outperform their peers and drive aggregate employment and productivity growth for the small business sector. Why do some MSEs expand rapidly, while others stagnate? What factors account for the wide variation observed in MSE growth trajectories? This paper explores a wide array of specific factors, but it is first useful to consider a broader conceptual frame work of opportunities and capabilities. Clearly, opportunities for profitable business activities shape the ability of an entrepreneur to expand his or her firm.

Yet, profitable business opportunities are a necessary but in sufficient condition for firm growth (Mead, interview, 2004). To take advantage of business opportunities, entrepreneurs must also possess appropriate capabilities, such as skills, resources, or technology. Most prominently, MSEs that demonstrate high, sustained growth rates are frequently termed "gazelles." These highly performing firms typically share two fundamental characteristics, they have profitable business opportunities and appropriate capabilities to harness these opportunities. Only a minority of firm s become "gazelles," which drive overall growth in the MSE sector. Some MSEs may face potentially lucrative business opportunities, but be unable to take full advantage of them due to in adequate capabilities. Although these "ponies" may expand quickly for short durations while trying to harness these opportunities, they often lack endurance as they do not have requisite capabilities for sustained growth. For instance, some small honey producers in Brazil initially experienced strong demand for their organic honey in open market places, and hoped to sustain growth by marketing to supermarkets. However, inadequate capabilities inhibited them from achieving this goal because inappropriate technology prevented them from satisfying the formal packaging requirements of supermarkets (Nichter, 200 4). In the interests of consistency and simplicity, this book defines MSEs as firms with up to 50 workers that are engaged in non-primary activities and sell at least 50 percent of their output. MSE owners and workers do tend to be disproportionately poor, with the incidence of poverty within MSEs typically higher than in medium-sized and large firms. On the other hand, MSEs may have substantial capabilities, but lack viable opportunities to capitalize on them. For example, a village tailors in India may have the ability and capacity to produce more saris, but faces weak local demand and has limited access to external markets. While such entrepreneurs' may experience lackluster growth for long periods, new opportunities (such as broader market access) can catalyze rapid expansion. Just like a "caterpillar" awaiting its metamorphosis, these MSEs exhibit substantial latent opportunities for growth. Many MSEs lack both profitable business opportunities and a host of capabilities such as skills, resources, and technology. These "tortoises" demonstrate the least proclivity to ward growth and their owners frequently focus instead on firm survival. Despite their lack of growth, these MSEs play a crucial role. In development, business growth should be considered within a broader perspective that includes income diversification and survival strategies.

Even "tortoises" without employment growth often provide essential sustenance for their owners and workers. In some cases, slow or nonexistent firm growth is attributable to the entrepreneurs' competing interests rather than a lack of capabilities. Women MSE owners, for example, may be especially likely to use business proceeds to purchase household necessities, to invest in parallel enterprises, or to assist their offspring in launching new firms. Such strategies are discussed later in the paper. This paper explores MSE growth by building on this conceptual framework of opportunities and capabilities. The macroeconomic context frequently also has indirect effects on MSE capabilities. For instance, when central banks raise interest rates to curb inflation, MSEs often becomes more credit constrained (even from inform al sources). Thus, the macroeconomic context has an indirect effect on MSE capabilities through another factor, access to finance. Similarly, education may not only directly enhance the capabilities of MSE owners, but also indirectly expand business opportunities through contacts and social networks. Both direct and indirect effects can substantially shape MSEs' growth trajectories. For example, a lead firm sourcing electronics parts from small suppliers can offer critical expertise and access to technology, while providing steady demand for MSE products. Value chains with certain characteristics, such as a non-hierarchical governance structure; can provide a context conducive to the development of strong vertical and horizontal linkages, thus enhancing MSE opportunities and capabilities.

16.2. Productivity and Growth

MSE growth defined simply as an increase in the number of firm employees may not be sustainable, but growth accompanied by improvements in productivity is likely to contribute to the desired development effect s discussed earlier, such as broad-based economic growth. As any businessperson will attest, firms exhibiting fast growth tend to be more productive than are their slow-growth competitors. Firm productivity refers to how much output a firm can produce with a given a level of inputs, and can be measured in various ways. Growth and Equity through Microenterprise Investments and Institutions (GEMINI) research found that relative to one-person firms, MSEs with two to five workers had substantially higher efficiency (interms of returns per hour of family labor), and those with six to nine workers had even higher efficiency (Mead and Liedholm, 1998). The studies suggested that the productivity costs of being small were attainable, once the business crossed the one-worker threshold. Last, during the firm life cycle, changes in productivity tend to be lumpy. For firms of all sizes, adding one or more units of capital or labor may not always result in immediate productivity increases. For example, new investments can result in temporary excess capacity.

16.3. The Business Environment

Contextual factors play a major role in shaping the opportunities of MSEs in developing countries. Most obvious ly, the overall state of the economy directly influences the availability of profitable business opportunities. Growth opportunities within existing markets, as well as the prevalence of untapped, profitable market niches, wax and wane as the business cycle evolves. It is hardly a surprise, then, that MSEs tend to grow m ore quickly during period s of overall economic growth (Liedholm, 2002). There are, however, some important nuances in the relationship between MSE growth and the overall business cycle:

the overall MSE sector expands during economic downturns due to an increase in survivalist-type activities, although individual MSEs may stagnate or contract. Further, during severe economic crises MSEs may be more resilient than their larger counterparts. Macroeconomic and relative price volatility is also an important issue, as experience has shown in Latin America and Sub-Saharan Africa (Tybout, 2000; Berry, 2004). The International Finance Corporation (IFC) recently conducted surveys of more than 10,000 firms in 80 countries, finding that both inflation and the exchange rate tend to afflict MSEs more than larger firms (Schiffer and Weder, 2001).

Macroeconomic trends may also affect MSEs indirectly, for example, credit constraints can limit firm capabilities, as discussed in the previous section. The regulatory and institutional environment in developing countries, notoriously burdensome when compared with developed countries, frequently hampers small enterprise growth. Econometricanaly underscore how these challenges disproportionately harm smaller enterprises (Beck, 2004).

Typically, larger firms can bypass these duties by qualifying for investment promotions, and they may be preferred in allocations processes (Liedholm, 2001). In addition, special subsidies and trade protection may offer greater benefits to larger firms, who are often more capable of lobbying (Tybout, 2000). Smaller firms more frequently report government policies to be unpredictable, and this uncertainty may be yet another factor reducing growth-enabling investments (World Bank, 2005).

Ironically, government policies that actually aim to benefit MSEs may also suppress growth if they provide disincentives for employment expansion. For example, India offers attractive incentives to small enterprises, but by some accounts, these measures backfire because growth beyond a specified level entails losing valued benefits (Mitra and Pingali, 1999; Little,1987). For example, the manufacture of certain products in India is reserved for small firm, which reduces incentives for firm expansion (World Bank, 2005). Some owners even split up their MSEs into several enterprises in an effort to make them look smaller (Kashyap, 1988). While there appears to be consensus on what constitutes a negative business environment, much less has been written about what a positive one looks like. There are a few generally agreed upon characteristics. These include consistency, so that business owners know what to expect and can assess risks; a stable macroeconomic environment, again, knowing what to expect from the future can be as or more important than having a low inflation rate or favorable currency position today; the existence of mechanisms for contract enforcement and dispute resolution; an uninhibited flow of capital for foreign and domestic investment; supportive labor laws;

access to information, and investment in education and technology. Consider the value chain for apparel. By most accounts, the apparel value chain includes five main parts:

- The supply of raw materials (e.g., fibers);
- The provision of components (e.g., yarn and fabrics);
- Production networks made up of garment factories;
- Export channels established b y trade intermediaries; and
- Marketing networks at the retail level (Greif and Memoedovic, 2003).

Within a given value chain, MSEs may be involved in any number of activities, including primary production, assembly, and service provision. Value chain characteristics can hold implications for the type and nature of opportunities for MSE growth. These includes growth and demand, sector and industry, and the organization or governance structure of the value chain. Conventionnal wisdom tell us that chains need to grow quickly to provide Opportunities for business growth all around, while stagnant or shrinking chains are likely to exhibit trends such as consolidation and "weeding out" of less competitive firms. However, experience in development also shows that value chain growth may be necessary but not sufficient to guarantee MSE participation or to ensure that MSEs be befit from growth. The discussion below highlights characteristics in addition to value chain growth that appear critical to MSE development. In addition to the characteristics of demand, certain in sect oral tendencies have shown to favor the participation, and survival over time, of MSEs. One of the most enduring works on small enterprise development, Modern Small Industry in Developing Countries, by Staley and Morse (1965), identifies five characteristics of industry functions that bode well for small firm participation:

- Seasonal activities;
- Low capital requirements;
- Relative labor intensiveness;
- Non-repetitive production processes, and
- Small production volumes.

In agriculture, for example, high population density makes smallholder collaboration, and thus cultivation, more viable; also, specific crop strains may offer advantages to smallholders, depending on the inputs, land conditions, and other requirements associated with their cultivation. Also important is the organization of the value chain, especially inter-firm relationships and power dynamics. Models described in the literature range from buyer driven (top-down and not very conducive to MSE participation) to producer-driven (usually seen in high-tech chains where producers are relatively sophisticated) to network or non-hierarchy cal chains where power relationships are more balanced. These (network) chains may hold the most promise for MSE participation and growths because they allow for inter -firm cooperation.

16.4. Inter-Firm Cooperation

Firms that participate in value chains (all, with the possible exception of subsistence agriculture) must, by definition, interact with other firms and related entities such as universities and regulatory agencies. This section explores the concept of inter-firm cooperation as a driver for MSE growth, with emphasis on three aspects: vertical linkages, horizontal linkages, and supporting markets. Individual firms for m commercial relationships, referred to in this paper as vertical linkages, with their buyers and suppliers. In some cases, similar firms may group themselves or be organized by an outside party to work together, this are referred to as horizontal linkages. Supporting markets are also important in value chains, for services such as finance; consulting, legal, and tax advice; market information; and skills training.

- *Vertical linkages*: A number of mechanisms are used to form vertical linkages, ranging from loose and informal understandings to rigid buyer-supplier agreements known as subcontracting 10 or outsourcing. Contracts may be exclusive and legally binding, and include precise specifications regarding price, quantity, delivery dates, raw materials, and production processes. Vertical linkages can facilitate MSE growth by expanding a firm's set of viable business opportunities and improving firm capabilities. Agreements with buyers can decrease the risks and costs associated with entering new markets by providing a guaranteed flow of orders, critical information about market requirements, and in some cases, reducing the need for capital investments (Aw, 2001).
- *Horizontal linkages:* Among the many legal and organizational options to institutionalize horizontal cooperation are cooperatives, associations, consortia, producer groups, and other collaborative structures. Horizontal linkages can help MSEs overcome many of the disadvantages of being small, for example by providing a way to consolidate production, improve their negotiating position with buyers or suppliers, access market information or services, or lobby for political or regulatory changes. Geographic and sectoral agglome rations of enterprises, or clusters, are

also a vehicle through which MSEs may participate in rapid growth. Clusters inevitably involve external economies: one firm's investments spill over to other firms in the cluster (Schmitz, 1999).

In addition, clusters may involve consciously pursued joint action, such as sharing machinery or developing a product together. In cases where clustered firms seek to serve the same market, competition, as well as cooperation, can drive innovation critical to competitive performance. The mere presence of clusters does not guarantee dynamic growth for MSEs (McCormick, 1998). The advantages from participation in clusters relate directly to the strength of vertical and horizontal linkages, and relationships with supporting organizations including trade associations, universities and vocational schools, financial institutions, and local and national-level government agencies. These actors, in addition to private business service providers, are key players in supporting markets.

- *Supporting markets:* Services provided through supporting markets are often directly related to improvements in capacity. For example, skills training may allow firms to offer new products, while finance may allow them to produce greater volumes. Access to market information or new technologies, on the other hand, may help firms respond to new opportunities. While topics such as access to finance and technology are treated below in the section on firm-level characteristics, it is useful to recognize that supporting services may be offered directly to MSEs on a fee-for- service basis or embedded in firm relationships, that is, delivered through vertical or horizontal linkages. Value chain relationships, including horizontal and vertical linkages, and connections to supporting markets, offer many tangible, indispensable advantages to MSEs. These advantages include increased production capacity and ways to consolidate production; increased efficiencies; mechanisms to spread both costs and risks; increased bargaining power for inputs or raw materials; and channels to obtain information about and improve techniques used for production, marketing, transportation, and technology. These relationships, then, play a critical role in facilitating MSE growth. Linkages can expand business opportunities and enhance fir m capabilities at the same time.

Chapter 17

Unemployment Issues and Steps Taken by the Government

17.1. The Nigerian Unemployment Milieu: Control Measures taken by the Government

At this juncture, the little efforts that have been made by the government to control unemployment to an extent shall be examined. Records from the Central Bank of Nigeria (1987), show the following efforts made by the government:

1. Youths Empowerment/ employment and skills programme. This was done in terms of provided minor jobs or 'white-colar' jobs.
2. Agricultural programmes such as the 'Operation Feed the Nation' by the Obasanjo Military Administration.
3. Provision of small scale industries and graduates employment.
4. Special public works (which were temporary). Some of these efforts are still in place while others are forgotten. For example, on the pages of the Businessday Newspaper, on August 4th (pg 13), 12th (pg 49), 21st and 27th, 2009 were the following headings revealed:
   - Intels' running of Onne Port Complex creates 8,000 jobs. (*NB*: Intels is Integrated Logistics Services, a private company in Port Harcourt working with some oil Companies).
   - Lagos Government creates 156,000 jobs in 6 months. (*NB*: The same government also provides temporary job opportunities with manageable earnings to secondary school students during long vacations as well as vocational and skill training).
   - Edo State Government lifts embargo on employment. Adding l additional to the grade 16 ceiling, making it 17 for employment opportunities.

- IITA creates 6,500 jobs through cassava project. (*NB*: IITA means International Institute of Tropical Agriculture).

The Federal Government too is not left out. It creates jobs for graduates and provides vocational and skill training for them. Sectors such as communication, transportation, oil, *etc*. both private and public are not left out in jobs creation.

With these facts above, one is likely to think that enough has been done. What? With more than 140 million populace and over 50 million labour force consisting of greater percentage of the unemployed, a discouraging figure? Much still needed to be accomplished by the government and all concerned individuals to provide jobs for the millions of Nigerian unemployed. And these provisions must spread across the nation, reaching both rural and urban areas.

17.1.1 Defining poverty

The concept of poverty is translated into policy through a more precise set of definitions and measures. While, as argued in the Introduction, it is important not to confuse definitions and measures, some of the issues raised straddle the two and loop back into conceptualizations. Definitions are the subject of this first chapter; measures are dealt with in the second. After a general discussion of different approaches to defining poverty, chapter 1 looks at the traditional opposition between 'absolute' and 'relative' definitions and at alternative formulations that attempt to reconcile the two.

17.1.2. Approaches to defining poverty

How we define poverty is critical to political, policy and academic debates about the concept. It is bound up with explanations and has implications for solutions. Value judgments are involved. Definition thus has to be understood as a political as well as a social scientific act and as such has often been the source of controversy. There is no single 'correct' definition. However, as we shall see, most researchers now accept that any definition has to be understood, at least in part, in relation to particular social, cultural and historical contexts? This has implications for studies that attempt to compare poverty in very different kinds of society. Definitions vary according to their narrowness or breadth, that is in terms of: whether they are confined to the material core; the nature of that material core; and whether they embrace also relational/symbolic factors associated with poverty, as identified in the Introduction. Nolan and Whelan are among those who argue for a definition towards the narrower end of the scale on the grounds that too broad a definition runs the danger of losing sight of the distinctive 'core notion of poverty' (1996). Following Townsend, they define poverty in terms of the inability to participate in society (which is broader than more 'absolute' definitions confined to subsistence needs), but emphasize that what is distinctive is the 'inability to participate owing to lack of resources' (1996: 188). This confines their definition 'to those areas of life where consumption or participation are deter- mined primarily by command over financial resources' (1996:193; Veit-Wilson, 1998, 2004). By implication they exclude non-material elements found in broad UN definitions, for example: 'lack of participation in decision-making', 'a violation of human dignity', 'powerlessness' and 'susceptibility to violence' (cited in Langmore, 2000). Similarly, they exclude some of the non-material aspects emphasized by people in poverty themselves, such as lack of voice, respect and self-esteem, isolation and humiliation (UKCAP, 1997; Galloway, 2002). Given that, as argued in the Introduction, the function of a definition is to differentiate the condition defined (poverty) from other conditions (non-poverty), it makes sense to pitch the *definition* of poverty towards the narrower end of the spectrum. Aspects such as 'lack of participation in decision-making', 'susceptibility to violence' and 'humiliation' are not unique to the condition of poverty; they are also associated with other conditions such as being Black in a White-dominated society.

However, in order not to lose sight of the condition's wider meanings and of the interpenetration of the material and the relational/symbolic; it is important that definitions of poverty are not divorced from wider *conceptualizations* such as that developed in subsequent chapters. Another source of variation in definitions of poverty, reflected in the literature on measurement, lies in whether they are rooted in conceptualizations that are concerned with, on the one hand, a person's material resources, especially income, and, on the other, with actual outcomes in terms of living standards and activities (Nolan and Whelan, 1996). As Stein Ringen puts it, 'in the first case, poverty is defined indirectly through the determinants of way of life, in the second case, directly by way of life' (1987). In practice, these two approaches are often treated as complementary (as in Nolan and Whelan's definition above and Townsend's below). Indeed, Ringen's own definition is not unusual in combining the two: 'a low standard of living, meaning deprivation in way of life because of insufficient resources to avoid such deprivation' (1987). Put simply, someone is ' "poor" when they have both a low standard of living and a low income' (Gordon *et al.*, 2000b). A.B. Atkinson makes a related, but more fundamental, distinction between a concern with standard of living and a concern with a citizen's '*right* to a minimum level of resources' (1989 emphasis added). The former is more common in the literature and as the basis for empirical research. The latter might be said to be implicit in measures of poverty based on the numbers falling below a certain point in the income scale or the level of income provided by a country's social assistance scheme .While the 'right to a minimum level of resources' has not been widely adopted as an explicit definition of poverty, it does have a value as one element in a broader conceptualization of poverty. It means that people 'are entitled, as citizens, to a minimum income, the disposal of which is a matter for them' and which 'may be regarded as a pre-requisite for participation in a particular society, as a guarantee of "positive freedom" (A.B. Atkinson,1990).

The conceptualization of poverty in this way is also helpful from the perspective of understanding and combating women's poverty. Following Atkinson, Stephen Jenkins suggests that a feminist concept of poverty can be described in terms of an '*individual* right to a minimum degree of potential economic independence' (1991). Although the feminist definition pro- pounded by Millar and Glendinning is not couched in the language of rights, it focuses on the individual's capacity to be self-supporting on the grounds that 'people who are financially dependent upon others must be considered vulnerable to poverty' (1992). The notion of vulnerability is helpful to understanding the situation of women without an independent income who nevertheless enjoy a comfortable standard of living.

17.2. Absolute' and 'Relative' poverty

This distinction has been central to post-war debates about how to define poverty. Definitions deployed in the late nineteenth and early twentieth century by Charles Booth and Seebohm Row tree, the pioneers of modern poverty research, were supposedly 'absolute' in the sense that poverty was said to be understood as lacking sufficient money to meet basic physical needs. At its most basic, absolute poverty is defined in terms of survival; more commonly it refers to subsistence, linked to a basic standard of physical capacity necessary for production (paid work) and reproduction (the bearing and nurturing of children). Nutrition is central to such definitions:' an absolute standard means one defined by reference to the actual needs of the poor and not by reference to the expenditure of those who are not poor. A family is poor if it cannot afford to eat' (Joseph and Sumption, 1979). Implicit in this statement is a rejection of the alternative 'relative' definition, developed by Townsend and articulated most fully in his monumental *Poverty in the United Kingdom.* Townsend criticizes the narrow subsistence notion of needs, divorced from their social context, upon which absolute definitions of poverty were based. According to his alternative, relative, definition: Individuals, families and groups in the population can be said to be in poverty when they lack the resources to obtain the types of dict, participate in the activities and have the living conditions and amenities which are customary, or are at least widely encouraged or approved, in the societies to which they belong. Their resources are so seriously below those commanded by the average individual or family that they are, in effect, excluded from ordinary living patterns and activities (1979).

The European Commission's definition, adopted in 1984, is similar in tone: 'The poor shall be taken to mean persons, families and groups of persons whose resources (material, cultural and social) are so limited as to exclude them from the minimum acceptable way of life in the Member State in which they live.' Unlike Townsend's, though, it does not spell out explicitly the dimension of *social participation*, which is key to the concept of 'relative deprivation' upon which his definition of poverty is built. Relative deprivation occurs when people 'cannot obtain, at all or sufficiently, the condition of life, that is, the diets, amenities, standards and services, which allow them to play the roles, participate in the relationships and follow the customary behaviour which is expected of them by virtue of their membership of society' (Townsend, 1993). The notion of 'relative', which links the concepts of relative poverty and deprivation, needs unpacking, as it embraces a number of different meanings. Broadly these fall into two, interrelated, categories concerning, first, the nature of the *comparisons* to be made in judging whether poverty exists and, second, the nature of human *needs*. We will look at each in turn. The essence of the *comparative* element of relative poverty lies in the idea that it is only possible to judge whether or not someone is in relative poverty in relation to other people living in the same society at the same point in history. This breaks down into three elements: the historical, the cross-national and the intra-national. It is common for those who lived through the terrible hardship of 1930s Britain to claim that 'real' poverty no longer exists. From the perspective of an understanding of poverty as relative, such a comparison is misplaced for it says nothing about what people need to live a decent life in the early twenty-first century. Within shorter time-scales, at a time of a steady rise in general living standards and rapid technological change' we constantly manufacture new forms of poverty as we drive forward the living standards of the majority without thinking what we are doing to those who cannot keep pace'.

Another common response to claims that poverty is prevalent in the rich countries of the North is to point to the deep poverty experienced by millions in the South. Again, the comparison is unhelpful to an understanding of what poverty means for those who live in the former and who face the demands, expectations and costs of living in an affluent consumer society. However, it does open up wider questions of global comparators between the South and the North. In a globalized world, in which countries are more closely bound together through trading systems, electronic communications and cultural networks and diasporas, it may increasingly become 'necessary to accept the relativity of need to the world's as well as to national resources' (Townsend, 1987). This places the issue of poverty in the South firmly in the context of global inequalities. The final comparative element of relative poverty places poverty in the context of inequality within *societies*. The act of comparison be it between those on lower and higher incomes, women and men, minority and majority ethnic groups, inevitably highlights any inequalities of material resources that may exist between the groups being compared. This is not, however, to say that relative poverty and inequality are synonymous, as is sometimes claimed. Inequality is concerned solely with the comparison between groups. Relative poverty adds to that comparison the notion of the incapacity to meet certain needs, broadly defined to include participation in society. It is logically conceivable, though unlikely, that a society could be very unequal but without poverty, if all its members had the resources necessary to participate fully in that society.

17.3. The Reconciliation of Absolute and Relative

Having seen how even supposedly absolute definitions of poverty involve an element of relativity we turn to the other source of challenge to the absolute relative dichotomy. This lies not in a denial of absolute poverty but in a reformulation of its relationship to relative poverty so as to integrate the two into one framework in place of two competing definitions. Such a framework, it is argued, can then be applied to both richer and poorer countries. Of particular significance here is Sen's work, discussed earlier, and its deployment by Doyal and Gough in the construction of their influential theory of human needs. An alternative attempt at combining absolute and relative within one framework can be found in the Copenhagen Declaration on Social Development, agreed at the 1995 UN World Summit on Social Development.

Chapter 18

Past Policies of Government – A Survey

18.1. Basic for Government Intervention in the Economy

One of the most powerful arguments to justify government intervention in the economy in the economy is the so-called concept of market failure. In a typical free enterprise system where demand and supply forces operate, the efficiency which results only relates to economic welfare. The social cost implications of economic activities tend to be ignored. Governments therefore intervene in the economy with the social welfare objective in mind. One popular way by which governments have intervened in the economy is through ownership of particular industries or outright nationalization of some private enterprises. The advantages usually claimed for such an action are as follows:

i. *The economics of scale argument:* It is argued that when there are a number of firms producing a particular product, the firms do not work their plants at the optimum level. In additional, there are duplication of efforts in the areas of research and promotion. If a monopoly power is, however, vested in a state enterprise, commercial, financial and technical economies of scale can be obtained.

ii. *The huge capital outlay argument:* It is also argued that there are certain projects which, because of the huge outlay and risk involved, are best handled by the government. Private firms, it is argued, may not be interested in investing in such projects even if they have beneficial effects for the economy as a whole.

iii. *The essential services argument:* There are certain services which can only be provided by a monopoly organization under state control. Such service included water supply, electricity provision; etc. it is therefore argued that the best way to control such a monopoly organization is for the state to own it. It is then that there can be assurance that such an organization will work in the public interest.

iv. *Increased productivity argument:* When the state controls an enterprise, workers tend to enjoy better working conditions since it is difficult to imagine that the government would want to exploit them. This can, in turn, encourage them to increase productively. This contrast sharply with a private enterprises situation where workers feel exploited, and as a result, are not motivated to give their best.

v. *The key industries argument:* It is also argued that some industries are of such strategic important to a country that private firms whose sole objective is profit, should not be allowed to go into such industries. These industries include, in the case of Nigeria, weapon producing industries, iron and steel industries and crude oil production. These industries, it is argued, should be run by the state for the general interest of the country, rather than leave them to provide firms.

vi. *Fiscal and incomes policy argument:* With the government itself owing some industries, it becomes relatively easier to use these industries to carry out some of the government fiscal policy and income policy measures. It tends to be more difficult if it were private enterprises that were involved, especially where such policies were not favourable to them. Opposed to these argument justifying state involvement in business operations, are the following arguments justifying privatization. That is, that the government should refrain from directly operating business enterprises.

It is argued that since the performance of public enterprises is not usually based on profit level, workers are not motivated to do their best. They are generally apathetic towards the services that are supposed to provide for members of the public. Decision making tends to be slow and bureaucratic which lead to general inefficiency. To buttress this point, the protagonists of this argument point to the appalling service rendered by some public enterprises such as N.E.P.A. Nigerian Airways and the Nigerian Railway Corporation. Lack of encouragement for individual initiative. Secondly, it is argued that public enterprises do not allow for individual initiative since most of what needs to be done remains in the "hands" of the government. Thirdly, it is argued that in an economy, especially a developing one like that of Nigeria, there is speed to attract foreign investment to speed up the pace of economic development. But foreign investors are not usually enthusiastic about investing in an economy in which there is too much governmental intervention. Foreign aid/loan issue a related argument to the above as regards public enterprises is the general unwillingness of international lending institutions to provide loans to countries in which government intervention appears to be too much. Indeed, one of the conditions insisted upon by the international monetary fund (IMF) for its much debated loan to Nigeria between 1984 and 1985, was that the government must divest itself of its holdings in a number of public enterprises. Apart from outright ownership of enterprises and nationalization, there are other ways by which government intervenes in the economy in order to move it in a desired direction. These other ways may include

a. The use of fiscal policy measures,

b. The use of monetary policy measures,

c. Incomes and prices guidelines.

These measures, however, shall be explained in the relevant sections of the syllabus to which they relate. The Nigeria enterprises promotion decree was promulgated in February 1972. The decree reserved 22 types of enterprises exclusively for Nigerians. The enterprises included retail trade (except for department store and supermarket), newspaper printing, tyre retreading, rice milling, advertising, bread and cake making, road haulage, bus services and tax is. In addition, aliens were prevented from being owners or part-owners of a further 33 types of enterprises.

If:

(i) The paid-up share capital did not exceed ₦ 200,000 or if the turnover did not exceed ₦ 500,000 (according to which of these two the Nigerian enterprises promotion board considered more appropriate). or

(ii) The paid-up share capital exceeded ₦ 200,000 or the turnover exceeded ₦ 500, 000 and equity participation by Nigerian citizens or associations was less than 40 per cent. The decree gave the NEPB tough powers to affect the sale of all businesses whose owners had not made genuine efforts to sell by March 31, 1974 and prescribed severe punishment for any Nigerian who acted as a "front" for a foreign-owned business. In November 1975, a panel was set up to examine the impact of the 1972 decree. As a result of its report, a second round of indigenization was unveiled in a broadcast by the head of state on June 20, 1976. The second round took effect in two stages. The first involved only the banks, which were required to increase their Nigerian equity from the 40 per cent level set in 1972, to 60 per cent by the end of September 1976. The second stage came into effect in January 1977, but its provisions were staggered, with the largest companies having to comply by June 1977 and the remainder by the end of 1978. The decree classified all enterprises into three schedules. Those in schedule I must be wholly owned by Nigerians. Those in schedule II must have 60 per cent Nigerian ownership, whilst those schedule III, the minimum Nigeria equity holding was 40 per cent.

The major expatriate trading companies like UAC, John Holt, Leventis, Paterson Zochonis and UTC avoided inclusion in schedule I (which covers much of the retail sector), by an exemption that transfer to schedules II, companies that would otherwise fall in schedule I, but operate in at least ten states and have an annual turnover of at least ₦ 25 million.

Chapter 19

Issues and Challenges in African Youths

19.1. Remedies to Combat Unemployment in Nigeria

(a) Rural development

One way of stemming the rural-urban drift, and thus control the resulting unemployment in our urban centers, it for the government to develop the rural areas. Not only should amenities be provided, but jobs should be created in the rural areas to stem the exodus. Fortunately, the Babangida administration seemed to have realized this with the huge amount allocation for the development of rural areas in the 1986 and 1987 budgets.

(b) Population control

Government must intensity the effort to control population growth Nigeria already has a huge population and with an estimated growth rate of about 3 percent annually, this can only compound the unemployment in the face of inadequate resources. It is therefore advocated that the government should renew the family planning campaign to check the population growth.

(c) Labour mobility

A situation where Nigerian prefers to remain unemployed, rather than go to work in other states for fear of discrimination, should attract government attention. All forms of discrimination – whether real or apparent, should be removed to make Nigerians accept jobs in states which are not their own.

(d) Labour intensive technology

Given Nigeria's huge population, a labour intensive technology should be preferred and encouraged rather than a capital intensive technology. This will make it possible to absorb a significant portion of the population into the labour force.

(e) An appraisal of education system

An appraisal of our educational system should be undertaken. More emphasis should be placed on technical and vocational discipline. The government seemed to have realized this with the establishment of the 6-3-3-4 system and the establishment of university of technology. Hopefully, our institutions should now be able to train technically skilled graduate rather than the liberal arts ones, who are not able to fill the vacancies in the job market.

(f) Agricultural should be encouraged

The growing number of unemployed people in the urban centers should be encouraged to take to farming. Loans should be provided by the banks for the acquisition of farm implements on easy terms to attract young energetic people into agriculture. The gains of this policy are twofold. The first is that it helps to increase food production. The second is that is that it helps in reducing the unemployment level.

19.2. Economic Growth and Economic Development

Economic growth is measured by increases in the real Gross National Product (GNP) per head of the population. It is a measure of the extent of increases in national income per head after adjusting for increases in the price level. Thus, if in 1988, the money national income was #40 million and in 1989, it was #80 million naira, this would suggest a 100 percent economic growth rate. But this would not necessarily be the case. If inflation rate in the economy between the two periods was say, 40 percent, then the real rate of growth is only 60 percent. Also although economic growth is possible under a situation of less than full employment, usually however, reference is made to growth after full employment level has been achieved. Economic growth is a necessary but not sufficient condition of economic development. Economic growth and economic development are closely related but they are not synonymous. Economic growth can exist independent of the state of economic development. Economic development is essentially investment in an economy; economic growth is increased production of an economy.

Economic growth is a term generally measured by the amount a country or region produces over a certain period of time. Economic growth is a narrower concept than economic development. It is an increase in a country's real level of national output which can be caused by an increase in the quality of resources, increase in the quantity of resources and improvements in technology or in another way an increase in the value of goods and services produced by every sector of the economy. Economic growth can be measured by an increase in a country's GDP (gross domestic product). Economic development of a country is defined as the development of the economic wealth of the country. Economic development is aimed at the overall well-being of the citizens of country, as they are the ultimate beneficiaries of the development of the country. The definition of economic development given by Michael Todaro is an increase in living standards, improvement in self-esteem needs and freedom from oppression as well as a greater choice. The most accurate method of measuring development is the Human Development Index, HDI, which takes into account the literacy rates and life expectancy which affects productivity and could lead to economic growth. It also leads to the creation of more opportunities in the sectors of education, healthcare, employment and the conservation of the environment. Gallopin (1994), emphasized that development is not synonymous with economic growth. Development according to him involves qualitative transformation, while growth is essentially quantitative increase. The goal of development is to increase the quality of life of the human population; economic growth is only one mean to this end. Furthermore, quality of life embodies the satisfaction of material and non-material human needs and the fulfillment of human desires and aspiration.

Many writers argue that, after decades of the so called development project aimed at raising the Third World out of poverty and improving the well-being of its citizens, one can speak of anything but true transformation. Yet thousands of children still die of malnutrition-related diseases every day, and millions of people still do not have access to clean water. Inequalities within countries themselves persist. (Katarzyna Karpowicz, 2008). Development has long been a catch-all phrase and it now seems that everybody has something to say on the subject. It is indeed crucial to know what one is talking about, if only to allow for the differentiation of development from economic growth with which it is equated all too often. In point of fact, it is believed that the apparent failure of the development project can be attributed to precisely this misunderstanding surrounding the notion of development itself. In the words of Seabrook (1993: 7): "All over the world, more and more people are being disadvantaged by a version of development which, even if it creates wealth, leaves them with a sense of loss and impoverishment". Thus, based largely on the comparison of two indicators used by major international bodies dealing with development – Gross Domestic Product (GDP) and Human Development Index (HDI), it is argued that growth and development are, actually, two different phenomena. In fact, economic growth is but a step in the direction towards development – one of major significance, indeed a precondition to it, but by no means can it be conceptualized as development itself. For a country to be generally recognized as a developed one, it also needs to be able to provide its citizens with basic resources and services, such as food, healthcare and schooling.

The problem with seeing development purely in terms of a country's economic performance is clearly visible if one considers the method used for measuring it. For a long time, the indicator used for measuring development was Gross Domestic Product (GDP) expressed in per capita terms. Those who support this view of development would have it that unless a rise in GDP per capita takes place, no development whatsoever is bound to happen (Storey 2003: 29). What follows is that the increase in GDP per capita is a sine qua non of development. True as it may be the fact that GDP is a purely economic indicator and, as such, does not allow for a change in social, cultural or political conditions. Also, it has been blamed for its falling short of providing a true image of the country's development, or the lack of it. Another major problem with GDP, and thus with the perception of economic growth as equal to development, is that it gives no indication of how fairly resources are distributed in the country. If development cannot solely be based on economic growth, as reflected in the increase in GDP, then what else does it encompass? Having stated that economic growth is but one of many elements of development, one may then try to look at some of the other factors that would be included in the definitions of development. In doing so, we may want to consider the UN's Human Development Index (HDI). HDI has been designed to remedy the shortcomings of GDP as a measurement of the progress achieved by developing countries. It not only allows for the growth in income per capita, but comprises also two other elements crucial to human well-being: life expectancy and educational attainment. Thus, development is seen as a "process of enlarging people's choices" in order for the people to "lead long and healthy lives, to be knowledgeable and to have a decent standard of living" (Human Development Reports.) Therefore, HDI is a comparative measure of life expectancy, literacy, education and quality of life. It is a standard of measuring well-being, and is used to distinguish whether a country is a developed, developing or an underdeveloped country. HDI was introduced as an alternative to conventional measure of national development such as level of income and rate of economic growth. Nigeria HDI is 0.459, which gave the country a rank of 156 out of 187 countries with comparable data.

19.2.1. Features of developing economy

We may now identify some features of an underdeveloped economy from that of developed one as follows:

1. In an underdeveloped economy, the fundamental problem is that of output, real income or the standard of living, as these economies are characterized by low productivity, low income and a poor standard of living. A vast majority are undernourished and without adequate shelter. On the other

hand, most of the developed countries at present enjoy a high rate of mass-consumption. In this economies, per capital real income has risen to a level at which a large number of people can afford consumption transcending food, shelter and clothing.

2. An underdeveloped economy, compared with an advanced economy, is under equipped with capital in relation to its population and natural resources. The rate of growth of employment and investment in such an economy lags behind the rate of growth of population. The resources are not only employed but also underemployed.
3. High rate of growth of population is an important characteristic of most of the underdeveloped economies. Population growth in underdeveloped countries neutralizes economic growth.
4. The central problem of underdeveloped economies is the prevalence of mass poverty which is the cause as well as the consequence of their low level of development. Shortage and scarcity are the main economic problems in these economies, whereas the affluent societies of advanced countries have economic problems resulting from abundance.
5. There is chronic unemployment in an underdeveloped economy. An advanced economy may have unemployment occasionally due to business fluctuations and a low marginal propensity to consume. Whereas an under developed economy is confronted with the problem of disguised unemployment in the sense that even with unchanged techniques in agriculture could be removed without reducing agricultural output.
6. Poor countries are poor in technology; advanced countries are advanced in technology. In fact, the level of technology attained in production is a reliable indication of the level of economic development. Employment of advanced technology goes along with large capital resources, high attainments in the fields of scientific research, greater availability of entrepreneurial skill and a good supply of efficient skilled labour. Thus, development of technology is the basic objective of the backward economy.

19.2.2. Development planning and Nigeria's economic reform programmes

(a) National Development Plans (NDP)

After independence in 1960 the need for a more formal and coordinated economic development resulted in the 1962-68 plan, which represented the first "national" development plan because of the extensive involvement of Nigerians in its formulation. One notable characteristic of the plan was the extent of coordination between the central government and the federating units. The Second National Development Plan, (1970-74), was more comprehensive and ambitious as crude oil production had already emerged as a major contributor to national income. The major focus of the 1970-74 Plan was the reconstruction of the war-torn economy. The objectives espoused were to establish a united, strong and self-reliant nation, develop a great and dynamic economy, provide the basis for a just and egalitarian society, transform Nigeria into a land of bright and full opportunities for all citizens and develop a free and democratic society.

Subscquently, two more development plans were initiated i.e. The Third National Development Plan (1975-80), and the Fourth National Development Plan (1981-85). The third plan involved extensive consultation which the private sector and the academia, and because Nigeria witnessed unprecedented oil boom in the mid 1970s, it benefited from the increased earnings from oil production. The main objectives did not diverge much from that of the earlier plan, though there was an added objective, namely, "indigenization of economic activities". The 1981-85 Plan, which succeeded it, was not remarkably different in objectives, except that it was introduced under a democratic administration. From the early part of the fourth plan period, Nigeria witnessed a lot of economic problems owing to excessive external reserve loss arising from capital flight, worsening terms of trade and drastic fall in domestic production. The economy was not performing at its optimal level. Rationing and price control were rampant, and as it the case with planned economies, there was always a shortage of some items and a surplus of others. In the midst of the problem, the plan was virtually abandoned. On top of the economic problems was a major political crisis, which culminated in the overthrow of the civilian government. The new military government promised accountability and was able to achieve some level of internal and external balance by reducing public spending and foreign exchange commitments through further import restrictions.

(b) Structural Adjustment Programmes (SAP)

The emergency of economic crises in the Nigeria society was the main cause of instituting the structural adjustment programme (SAP). To avert or moderate the effects of the economic crises, the government towards the middle of 1986 announced the commencement of SAP for the period July, 1986 to June 1988. The aim is to effectively alter and restructure the consumption and production patterns of the economy as well as eliminate price distortions and dependence on oil revenue and import of consumer and producer goods which then characterize the economy. In addition, SAP is expected to enhance the non-oil export, especially the tradable agricultural products thereby improving the economy and bringing the economy back on the path of steady and balance growth. The main strategies of SAP include: institutional reform; pricing policy and specific production scheme for agricultural crops. Economic reform in Nigeria in the 1990s witnessed a review of SAP because of its limited impact.

(c) Review of National Economic Empowerment and Development Strategy (NEEDS)

However, the advent of democratic rule in 1999 brought about a rethinking of economic reform in Nigeria. The widespread disillusionment arising from the failure of previous reforms and the experiences gained from the participatory approaches employed in the development of the interim Poverty Reduction Strategic Paper (PRSP) are the catalysts of a new social and economic development framework termed "National Economic Empowerment and Development Strategy (NEEDS). NEEDS provided the basis and philosophical undertone for economic reform. It has four main strategic features that determine the nature and scope of economic reform measures:

- Reforming the way government works and its institutions;
- Growing the private sector;
- Implementing a social charter for the people; and
- Re-orientation of the people with an enduring African value system.

The NEED document involved a participatory process, with the State and local government levels keying into the programme through their own versions- the State Economic Empowerment and Development Strategy (SEEDS) and the Local Economic Empowerment and Development Strategy (LEEDS). These comprised broad agendas of social and economic reforms based on four key strategies, including private sector participation as the main driver of the economy; implementation of a social charter for improved security and welfare; value re-orientation; and focus on attracting Foreign Direct Investment (FDI) (OSGF, 2012).

(d) Vision 20:2020

In 2007, Vision 20:2020 was unveiled by the government, declaring the intention of catapulting Nigeria into the group of twenty top economies by the year 2020. The vision was anchored on the Seven Point Agenda (SPA) of Yar'Adua/Goodluck administration, with the objectives of meeting the energy, transportation, human capital development, land tenure, food security needs of Nigerians, among others.

(e) Transformation Agenda.

There is indeed a great concern about the failure of previous strategies to achieve the desired effects on the development of Nigeria. Probable ineffectiveness in policy implementation inevitably caused the failure of previous efforts at repositioning Nigeria. The emergence of President Goodluck Jonathan in 2010/2011 set the background for a new order—a search for a strategy for the recreation of Nigeria. This, undoubtedly, necessitated the introduction of the Transformation Agenda. (OSGF, 2012)According to the Office of the Secretary to the Government of the Federation (OSGF) (2012), the cardinal elements of the Transformation Agenda of the current administration include constitutional and electoral reforms and transformation of the budgeting process. Others will include overcoming the current security challenges through ensuring peaceful co-existence, tolerance, employment generation, poverty alleviation and job creation for our youths aside from determined focus on developing critical infrastructure and human capital development. There is an overriding push for effective planning, coordination and continuity of policies and programmes with creation of an enabling environment for private sector participation through Public Private Partnership (PPP). The Administration is also poised to deliver good governance through tackling corruption, value re-orientation, accountability and transparency as well as Monitoring and Evaluation for all priority policies, programmes and projects.

19.3 Youth Issues on Sexual Exploitation

Sexual exploitation is a fundamental violation of children's rights. It is an attack on human dignity and inhibits the social and economic development of a country: destroying a child's life through sexual exploitation also destroys his or her chances of integrating into society. The Stockholm First World Congress (1996) against Commercial Sexual Exploitation of Children, and the Yokohama Congress in 2001, helped to raise awareness of the scale of the phenomenon, which is based on three closely linked themes:

- Sexual abuse occurs when a child is used for sexual purposes. In order to determine whether behaviour is abusive, it is important to consider certain criteria and the circumstances in which the act occurs. Sexual abuse is carried out by a person whom the child trusts, and could be a parent, brother, a member of the extended family, a teacher, a stranger or any other person in a position of power, authority or control over the child;
- Sexual violence is characterized by any type of sexual relations that are imposed by force, coercion, threat or surprise;
- Commercial sexual exploitation refers to the payment in cash or kind in exchange for sexual relations. In the region, it still occurs in an informal context and can appear in four forms: trafficking in children, prostitution, pornography and sexual tourism.

National situation analyses, specific studies carried out for several years by ministries, associations and multilateral agencies working on the sexual exploitation of children in West and Central Africa have all contributed to raising awareness of the characteristics and scale of this multiform phenomenon. This situation analysis of sexual abuse and exploitation of children in West and Central Africa was prepared as part of the regional technical meeting on sexual exploitation of children held in Dakar on 24-25 September 2008. The objectives of the report are to trace the development of the phenomenon, to present actions to combat the different forms of sexual exploitation of children, and to propose recommendations. The report is based primarily on a literature review, country reports and surveys from the 24 countries in the region. Since the Rabat Arab-African Forum (2004), the 24 countries1 that comprise West and Central Africa have carried out over a hundred studies and analyses concerning sexual abuse and exploitation of children. This research has improved understanding of how abuse and exploitation can occur, as well as the determining factors and developing trends:

- Commercial sexual exploitation of children in all its forms appears to be intensifying. There are many links between trafficking for sexual purposes, child prostitution, sexual tourism and child pornography.
- Trafficking in children for economic reasons coexists with trafficking in children for sexual purposes. The development of sexual tourism leads almost inevitably to the development of child prostitution and child pornography (certain abusers film their victims). Regions that are popular for sexual tourism are also very often favourite destinations for trafficking, both internally and externally, in children for sexual purposes in order to meet the sexual demands of tourists.
- Sexual abuse and exploitation in conflict and post-conflict situations have undergone a change: offenders are predominantly civilians that are known to the victims.
- Sexual abuse, exploitation and violence in schools remain a concern.
- Girl domestic workers are often victims of sexual abuse and exploitation.
- There is a significant relation between the sexual abuse and exploitation of children and the worsening HIV/AIDS pandemic.
- The rate of early and/or forced marriages remains relatively high in many countries in the region. They are:

  Benin, Burkina Faso, Cameroon, Cape Verde, Central African Republic, Chad, Congo, Côte d'Ivoire, Democratic Republicof the Congo, Equatorial

Guinea, Gabon, Gambia, Ghana, Guinea, Guinea-Bissau, Liberia, Mali, Mauritania, Niger, Nigeria, Sao Tome and Principe, Senegal, Sierra Leone and Togo.

- Sexual abuse and exploitation of children is a gender issue because girls are most affected. However, there are recorded cases of boys being victims of sexual abuse and commercial sexual exploitation (especially sexual tourism). In fact, certain foreign pedophiles (men and women) exclusively seek sexual relations with boys. The issue of boy prostitution remains taboo, particularly when it concerns homosexual relations. In certain countries where homosexuality is illegal, if a boy under 18 years old is raped, he can also be punished by law for having had homosexual relations.

In contrast, little information is available on sexual abuse, exploitation and violence against:

- Children within their family
- Street children and Islamic students (*talibés*)
- Children in institutions (orphanages, shelters and refuge centres)
- Children in conflict with the law, placed in detention
- Children living with a disability

The actual scale of all forms of sexual abuse and exploitation of children is difficult to determine for the following reasons:

- Low rate of reporting because victims fear reprisals, stigmatization and social exclusion;
- Persistent cultural resistance concerning sexuality and sexual violence;
- Cases are settled informally;
- No centralized system for collecting and processing data;
- No centralized system for monitoring and evaluating the situation of children and the impact of interventions;
- Weak judicial mechanisms and support services for victims that can guarantee protection and provide care;
- Impunity for certain offenders. After analysing the determining factors in the sexual abuse and exploitation of children, it would appear that certain social norms promote the persistence and intensification of particular forms of child sexual abuse and exploitation;

- Endemic poverty leads to worsening living conditions for children and make them more vulnerable to sexual abuse and exploitation;
- The HIV/AIDS epidemic is both a cause and consequence of sexual exploitation of children;
- Humanitarian crises stemming from conflicts or natural disasters increase children's vulnerability to sexual abuse and exploitation;
- The globalization of the sex industry has a strong impact on all forms of commercial sexual exploitation of children;
- Trafficking networks are increasingly organized;
- All these factors are correlated and influence each other.

Despite numerous efforts by many countries in the region to develop national action plans, implementation is very often partial or incomplete due to a lack of follow-up support. There are a great number of constraints:

- Weak institutional capacity among those responsible for devising, implementing and monitoring action plans: weak capacity in the development, programming, management and monitoring and evaluation of strategies and programmes;
- Inadequate human resources in terms of numbers and quality: high mobility, inappropriate profile/post, concentration of personnel in urban areas to the detriment of the rural zones;
- Low budget allocation that is dependent in part on development partners, poorly understood budgetary procedures, and tardy disbursement of allocated funds;
- Multiplication of national action plans (trafficking in children, orphans/ children vulnerable to HIV/AIDS, children in conflict with the law, etc.) can lead to overlap, and a lack of rationalization and optimization of available means;
- Inadequate or non-existent consultation and synergy between various ministries during development of plans;
- Inadequate or non-existent centralized information systems of monitoring and evaluation to measure developments in the situation of children.
- There have been a great number of legislative reforms. However, in spite of these efforts, in many of the countries in the region, there are still gaps in the law relating to:

i. The legal age of marriage;

ii. Certain types of sexual exploitation: child pornography, sexual tour ism and child prostitution;

iii. The principle of extraterritoriality;

iv. Protection and redress for children victims of sexual abuse and exploitation;

v. Free medical and legal aid for children.

The effective application of laws is a considerable challenge in all countries in the region as a result of the following constraints:

- Lack of formal, institutionalized complaints procedure;
- Weak judicial systems;
- Reluctance of families to get involved in legal proceedings;
- Persistent reliance on traditional "laws";
- Children and families unaware of their rights;
- Impunity.

Many awareness-raising campaigns involving the media, communities and children have been carried out. However, awareness is still fragmented, sectoral and often of limited duration; the messages are not always adapted to the target populations and their impact is not measured. Traditional behavior and practices related to child sexual exploitation still persist. Social protection mechanisms for the most vulnerable populations suffer from ineffective implementation. Social inequalities remain commonplace. Regarding the provision of specialized psycho-social care, reintegration and monitoring children victims of sexual abuse and exploitation, numerous actions have been carried out, most often by NGOs. In fact, to compensate for the inadequate responses from ministries, many associations and NGOs have created their own programmes and structures; however, their activities are geographically limited and their care provision and monitoring capacities are weak. In addition, the number of spaces, centers and programmes specialized in hosting, counseling and offering psycho-social support, as well as reintegration and monitoring children victims of sexual abuse and exploitation, also remain weak. Current capacity is already overextended. These structures are not subject to regulation and are not supervised by the relevant ministry responsible. They present a great many constraints:

- No standards for child care provision;
- Lack of a specialized approach to the specific problems related to the sexual exploitation of children;
- Lack of qualified human resources;
- Inadequate financial resources, limiting intervention capacities and jeopardizing the long-term sustainability of their actions;
- Insufficient alternatives to reintegration;
- Difficulties in obtaining reliable data due to the absence of a system for collecting and processing information;
- Haphazard child monitoring mechanisms;
- No means for measuring programme impact on the evolution of the situation of children. Despite greater visible participation by children, there is still room for progress as regards;
- Access to civil rights for children concerning access to information and the freedom of expression and association;
- Allowing child victims to participate and taking their opinions into account;
- The functioning of children's committees and parliaments;
- Skill in applying the participative approach by all actors involved with children.

Setting up control mechanisms for intersectoral and multispectral cooperation represents a major challenge for all countries in the region. The development of a global, integrated system of child protection against all forms of sexual abuse and exploitation is wholly dependent on the quality of cooperation. Indeed, establishing a realistic child protection system is dependent on particular constraints and risks related to:

- The commitment of all the actors and partners concerned in achieving the anticipated results;
- Synchronizing the different components of a response;
- The institutional capacity of the ministry responsible for coordinating implementation and monitoring and evaluation;
- Interministerial coordination with respect to priority goals;
- The time frames for introducing legislative and institutional measures that have been identified;

- The availability of required human resources;
- The allocated budget and the timing of its disbursement.

The national action plan is not simply a list of activities and programmes supported by various actors, but should correspond to the implementation of this system of protection.

19.4 Child Protection

Implementing such a system of protection is above all a process that is based on

- A complete adoption of the policy by all actors, not only in terms of content but also its principles and approaches;
- A programme that requires measurable results in terms of implementation of child rights interventions. The programme should ensure a rationalized and optimized use of resources; and
- Clarifying responsibilities of those in authority over children.

It is clear that the protection of children from all forms of sexual abuse and exploitation depends above all on the effective commitment of the State. The degree of interest by public authorities in child protection, their realization of the scale of the problem and their determination to protect all children without distinction are of vital importance.

Sexual exploitation is a fundamental violation of children's rights. It is an attack on human dignity and inhibits the social and economic development of a country: destroying a child's life through sexual exploitation also destroys his or her chances of integrating into society. Every study and survey regarding the sexual exploitation of children, whether for commercial reasons or not, indicates that it is both a cause and a consequence of other issues, such as poverty, changes in traditional modes of socialization and armed conflict. It is clear that a large majority of children victims of commercial sexual abuse, violence and exploitation are those who are already in vulnerable situation. Sexual exploitation of children is based on three closely linked themes:

- **Sexual abuse**: is the act in which a child is used for sexual purposes. The abuse is carried out by a person (adult or older child) whom the child trusts (parent, brother, a member of the extended family, teacher, guardian, etc.) or any other person in a position of power, authority or control over the child. The abuse may be physical, verbal or psycho-emotional. There is no informed consent from the child because of the pressure imposed by the adult, on whom the child is dependent or feels dependent: this pressure can cloud the child's judgment and restrict the independence of his or her responses. The pressure may be physical, but is most often psychological: seduction, praise, reward, blackmail and threats, as often about keeping the act a secret as about the act itself. The concept of sexual abuse is complex because it covers not only physical acts (molestation, rape), but also behaviour (harassment, attempted molestation or rape, sexual remarks) or even intentions (emotional or material blackmail, using deception).

- **Sexual violence** is characterized by any type of sexual relations imposed on children by force, coercion, threat or surprise.
- **Commercial sexual exploitation** is characterized by the payment in cash or in kind in exchange for sexual relations and implies the idea of profit, whether economic, social or other. The commercial sexual exploitation of children is defined as the sexual abuse of any person under 18 years old, for remuneration in cash or kind to the child or a third person or persons. The child is treated both as a sexual object and as a commercial object. The commercial sexual exploitation of children constitutes a form of coercion and violence against children, and amounts to forced labour and a contemporary form of slavery.

There are four main forms of commercial sexual exploitation: trafficking in children for sexual purposes, child prostitution, child pornography and sexual tourism involving children.

- Trafficking in children for sexual purposes refers to all acts related to the recruitment, transport, transfer, harbouring or receipt of children within a country or across borders, by means of deception, coercion or force, indenture by debt or fraud, for the purpose of putting children in a situation of sexual abuse or exploitation.
- Child prostitution means the use of a child for sexual activities for remuneration or other form of consideration.
- Convention on the Rights of the Child, Articles 34 and 35.
- Convention on the Rights of the Child, Articles 34 and 35; WHO World report on violence and health (2002).
- Optional Protocol on the sale of children, child prostitution and child pornography (Article 2); Stockholm Declaration; ILO Convention 182; Protocol to prevent, suppress and punish trafficking in persons.
- Child pornography refers to any representation, by whatever means, of a child engaged in real or simulated explicit sexual activities or any representation of the sexual parts of a child for primarily sexual purposes.

- Sexual tourism involving children is defined as the sexual exploitation of children by as person or persons travelling within their own country or travelling from their country to another country that is generally less developed, in order to indulge in sexual activities with children.

19.5 Child Pornography and Child Prostitution

The "Optional Protocol to the Convention on the Rights of the Child on the sale of children, child prostitution and child pornography" defines in Article 2(c) child pornography as "any representation, by whatever means, of a child engaged in real or simulated explicit sexual activities or any representation of the sexual parts of a child for primarily sexual purposes". In Cameroon, a study carried out by Association Enfant, Jeunes et Avenir in July 2006 (Mengue, 2006) highlights that adolescents aged between 14 and 20 years old are used in the production of pornographic materials; some victims are forced to participate, others receive remuneration. In the same report, certain neighbourhoods of Yaounde and Douala (Ekounou, Mvog-Ada, Pakita or "the crossroads of joy" and Mini Ferme Melen) have strip-tease clubs where boys and girls pose naked and are filmed. In Senegal, 1 girls are recruited for the production of pornographic films, sometimes with sexual relations with dogs. In Nigeria, girls from Benin are reported to be used in the production of pornographic films (ECPAT, 2007a). In Ghana, pornographic video cassettes featuring girls are reportedly in circulation (ECPAT, 2007d). In addition, the absence of supervisory measures results in a high exposure of children to pornography in many of the countries in the region:

- Visiting Cybercafés, where often very young children connect to pornographic sites;
- Projection of pornographic films that are accessible to minors;
- Rental of pornographic cassettes to minors.

Child prostitution means the use of a child for sexual activities for remuneration or other form of consideration. Based on certain reports and information produced by country surveys, it would appear that the prostitution of girls is continually expanding. In the Democratic Republic of the Congo, a study on the sexual exploitation of children in the mining areas of Katanga and Kasaïs reported that girls aged between 10 and 17 years old worked as prostitutes, either occasionally or permanently, in the neighboring villages, inhabited quarries or residential neighborhoods' (UNICEF DRC, 2008). In Cameroon, Kribi and Limbe have been identified as centres of child prostitution. In Senegal, prostitution of minors occurs on beaches, in hotels, bars, nightclubs and other tourist places. In a system of sponsorship and adoption for money or services for the family, the child is subjected to sexual relations, often forced, with the guardian. In Ghana, girl porters, "kayayes", aged between 12 and 13, work as prostitutes in the streets, barsand hotels. In Togo, a qualitative study of young female sex workers in Lomé prepared by PSI/Togo states that girls aged between 10 and 24 years old work as prostitutes in bars, restaurants, brothels or in other clandestine places. The Human Rights Watch report, "Borderline Slavery: Child Trafficking in Togo", reveals that some girls were forced into prostitution in Lomé, in the neighbourhood called, *marché du petit vagin* (literally, "Little Vagina Market"). In Burkina Faso, a survey of under-age female prostitutes in December 2005, carried out by the Department of Social Action and Education at Ouagadougou Central Town Hall, noted that girls under the age of 10 and living in the street or serving as a guide for beggars, are increasingly being driven into prostitution. In Mali, a rapid assessment carried out in 2005 on commercial sexual exploitation of children highlighted the scale of the prostitution of girls aged between 12 and 18 years old. Most of them were Malian, but some came from Burkina Faso, Cameroon, Côte d'Ivoire, Guinea, Nigeria and Sierra Leone. The prostitution takes place in the streets, in brothels and bars; in fact, several bars have rooms for this purpose. Some girls work during the day (street vendors) and work as prostitutes at night. The girls were introduced into sex work by a girl friend or other close friend or relative, or after having been tricked by adults who promised a better future. Some had suffered family pressures, and others were victims of trafficking or had fled armed conflict. In the Republic of the Congo, a report prepared by UNICEF in 2007 shows that female street vendors aged between 12 and 15 years old are increasingly becoming victim to prostitution organized in bars or *ngandas*, in hotels or brothels. The girls are illiterate or have very little schooling and fall into the hands of networks of pimps who promise them a well paying job, or a visa for Europe or South Africa. The girls work seven days a week, three weeks per month (one week per month they are considered unavailable because of menstruation). They are subjected to verbal, physical and sexual violence by their pimps, and their possessions are often confiscated. Child prostitution is very lucrative and brings in a lot of money for pimps, on average 3000 Francs CFA per trick. According to the report, the monthly turnover was estimated to be 2 100 000 Francs CFA (UNICEF, 2007d). In Gabon, a survey carried out by the Ministry of National Solidarity, Social Affairs, Welfare and Poverty Reduction (2007) indicated an increase in under-age female prostitutes in Port-Gentil. There is little documentation on the prostitution of boys because of the taboo surrounding homosexuality in many countries in the region. Prostitution is closely linked to sexual tourism and child trafficking, as evidenced in the presence of brothels near borders, and the high visibility of child prostitution in tourist areas. The development of sexual tourism in certain areas leads to a rise in child prostitution and child trafficking in order to satisfy the demand of sexual tourists. In addition, internal trafficking also encourages the development of prostitution.

19.6 Issues on African Youths

Despite these efforts, in many of the countries in the region there remain gaps in the law regarding

- The legal age of marriage;
- Certain forms of sexual exploitation: child pornography, sexual tourism and child prostitution;
- The principle of extraterritoriality;
- Measures of protection and compensation for children victims of sexual abuse and exploitation; free medical and legal aid for children.

In every country in the region, the effective application of law is still a problem: the number of children victims of sexual abuse and exploitation that have access to legal protection remains low, and a large number of offenders benefit from total impunity. There are many factors that explain the difficulties in applying the law.

- *Low rate of reporting:* The number of reported cases of sexual abuse and exploitation remains low because there are few or no support mechanisms that are easily accessible to children and that guarantee their protection. NGOs that offer support services for children rarely act as plaintiffs, and the children and their families are not well aware of the resources available (telephone helpline, counseling centre, etc.).
- *Access difficulties for children:* In order to begin legal proceedings against the offender and obtain compensation, the victim must press charges with the police. In some countries in the region, the minor must be accompanied by a relative or guardian. The immediate consequence is that children who are victims of sexual abuse within their family, or children without family, cannot press charges. A number of countries have set up procedures that allow the child to go to the police unaccompanied (Benin), or to call a free telephone number or make a simple oral or written accusation (Cameroon).

In order to facilitate procedure and to guarantee confidentiality in cases of sexual abuse and Exploitation at school, a system that allows anonymous accusations has been set up in some Countries (the Gambia, Ghana, Niger, Nigeria, Senegal and Sierra Leone).

- *Proving sexual abuse or exploitation: an additional difficulty*: It is up to the victim to provide proof of sexual abuse or exploitation by way of a medical or forensic certificate. However, victims are unaware of this obligation, and in any case, the expense of obtaining a certificate inhibits beginning legal proceedings. Victims do not have access to free medical and legal aid .In addition, the victim's age is of fundamental importance in order to qualify the offence; yet many children were not registered at birth and can therefore not proof their status as a minor.

- *Ignorance of the law:* The children and their families have limited or no knowledge of laws concerning child protection, and are therefore not even able to demand and defend their rights .Illiteracy means that people do not have direct access to understanding the law.

- *Significant lack of materials and qualified human resources:* Child protection units and juvenile courts do not exist in all countries, and there are strong disparities within a single country, which makes access to the police and legal services all the more difficult. Due to a lack of resources, police and legal services do not all have special places for children that can guarantee confidentiality and their protection during multiple hearings. The number of police officers and judges that have received special training in the protection of minors remains very low. The lack of resources also makes intersectional cooperation more difficult (police-justice health education).

Special training courses have taken place for all personnel involved in legal protection for Children, but there are sporadic and sectoral. To compensate for this deficiency; paralegals have been trained to work with communities that do not have close access to legal/police services. Informal settlements and compensation determined by the community. Families of child victims, especially in rural areas, tend to resort to:

- Settling cases informally by direct negotiation with the offender;

- The traditional chief, who will decide on the terms of the sanction against the offender and the traditional purification ceremonies;

- Sorcerers. The reasons for resorting to this type of settlement include the following:

- The shame and taboo that surround sexuality;

- The fear of reprisals and stigmatization;

- The influence of social norms;

- The lack of trust in justice (corruption, impunity, etc.). In this kind of settlement, the interests of the family and the community take precedence over that of the child.

- The slow speed and cost of legal proceedings drive families and victims to abandon their case or to withdraw.
- Sanctions are not always in accordance with the law due to the difficulty some judges have in qualifying the offence, given the multiplicity of concepts;
- Leniency in sentencing;
- Lack of deterrent measures;
- Pressure from offenders;
- The not-illegal status of certain forms of child sexual exploitation;
- Impunity of some offenders, thanks to their social status.

Compensation for harm done to the child is not systematic. When it occurs, it is often to the benefit of relatives or third parties. Little or no regular monitoring of places associated with child sexual exploitation. Although some countries have introduced monitoring mechanisms for bars, dance clubs and hotels, they are not used regularly. When police raids occur in these places, it is the child victims who are rounded up. Effective application of the law is a significant challenge in all countries in the region because of the following constraints:

- Lack of institutionalized system for accusations;
- Weak judicial systems;
- Reluctance of families to deal with legal issues;
- Continued reliance on traditional "law";
- Ignorance of children and their families of their rights;
- Impunity.

In terms of prevention, countries in West and Central Africa have made considerable efforts to:

- Raise awareness among populations of sexual abuse and violence against children;
- Raise awareness and inform populations about HIV/AIDS;
- Inform populations about the rights of the child and child protection laws;
- Change some social norms associated with sexual exploitation;
- Promote a culture based on the rights of the child.These campaigns consist of:

- Organizing sensitization days, educational discussions and seminars;
- Producing material such as posters, comic strips for children, handbooks, brochures, radio and television announcements, etc.

In the Democratic Republic of the Congo, a national campaign, "VDAY" is in preparation, with the theme, "Stop the rape of our most valuable resource: The rights of women and girls in DRC". A film on the life stories of victims of sexual violence in DRC will soon be shown in churches and schools. Ministerial circulars and orders have been introduced in many countries in the region with the aim of preventing sexual abuse and exploitation of children in schools. The media's involvement varies according to the country, but in general they play a significant role. The press denounces cases of child sexual exploitation and ensures the transmission of information to the general public. In Mauritania, radio and television programmes are presented by children; Cameroon broadcasts television debates; and the Central African Republic and Togo broadcast radio programmes concerning child sexual exploitation. In Benin, Senegal and Togo, networks of journalists for the protection of the rights of the child have been formed. Members are trained in the ethics and practical details of presenting information concerning children. Much effort has gone into mobilizing community involvement in order to gain a strong commitment to preventing and combating sexual abuse and exploitation of children. The results obtained by the NGO TOSTAN in Senegal clearly illustrate the importance of community involvement in preventing FGM and early marriages. The introduction of the "public statement", a heavily mediatized event that unites all communities, has led to the abandon of FGM and early and/or forced marriage in 3 140 communities in Senegal, that is, 62.8 percent of the 5000 communities where these practices occurred. This community-based initiative has also led to an improved awareness of HIV/AIDS prevention, systematic registration of births and encouraged school attendance for girls. However, it should be noted that the other forms of child sexual exploitation have not been targeted and that geographical coverage is not national.

In Burkina Faso, religious leaders and senior police personnel have been heavily involved in the fight against FGM. Village committees to combat sexual violence, trafficking and the worst forms of child labour have been set up in Côte d'Ivoire. As part of the combat against trafficking, many countries in the region have established village committees. In order to prevent sexual abuse, exploitation and violence against children at school, preventative measures that involve children have been introduced:

- Peer mediation (Guinea)
- Peer educators (Togo)
- Observatories on violence
- Children's government (Guinea)
- Life skills clubs to deal with HIV/AIDS (Sierra Leone)
- Student committees (Togo)
- Child-to-child information, education and communication activities (Benin)
- Children's governments within schools (Mali).

In the Gambia, the strategic framework for combating HIV/AIDS 2003-2008 includes a national school-based and community peer education and life skills programme. The objective is to sensitize 100 percent of youth aged between 15 and 24 by 2008. Six countries (Cameroon, the Gambia, Ghana, Niger and Senegal) have adopted a code of conduct for teaching and administrative staff in schools that explicitly prohibits sexual abuse, exploitation and violence. With the goal of preventing sexual abuse and exploitation of children in international organizations, Save the Children UK, in collaboration with ECPAT International and UNICEF, has developed a training kit on the security of children in organizations. Special attention is given to how aid workers from humanitarian agencies and peacekeeping troops respect the guidelines for the protection and response to sexual and gender-based violence against refugees, returnees and internally displaced persons (UNHCR, 2003). In contrast, the code of conduct for child protection against sexual exploitation in tourism does not appear to have been adopted by countries that are popular tourist destinations.

References

A Cemoghi, Deron, Jounson, Simon, Robinson, James, Thaica Uroen, Y. (2003). Institution caused, macro-economic symptom volatility 'crises and growth journal of monetary Economics 50, 49-123 (January).

A Look at Structural VARs, "Journal of Econometrics, Vol. 83, No. 1-2 (March-April), A Welfare-Based Approach," pp. 294–348, in Norman Loayza and Raimundo Sato.

A.J. Chhibber, et al., "Inflation Price Controls and Fiscal ey Supply and. 1, (2000), l. 21, No. 2, 1983, pp. 23-42. Omic Research, thesis, University of Ibadan, Nigeria, 1994.

A.L. Kilindo, "Fiscal Operations, Monpp. 1Inflation in Tanzania," African Economic Research Con-sortium, AERC Research Paper No. 65, Nairobi, 1997.

Abbott, J.C. (1986). Marketing Improvement in the Developing World: What Happens and What We Have Learned. FAO Economic and Social Development Series No. 37. Rome: Food and Agriculture Organization.

Abel, A. (1984). "Empirical Investment Equations: An Integrated Approach". Presented at Carnegie.

Abubakar, D. (2001). Globalization, social sciences and Nigeria in the 21st Century. The Nigerian Social Scientist, 4, 1. Activity, 1985(1):263-99.

Accessed April 08, (2014). http://www.theguardian.com/world/2014/feb/09/brazil-sex-tourism-world-cup.

Accessed April 20, (2014). http://www.unric.org/en/human-trafficking.

Accessed April 25, (2014). http://www.palantir.com/2013/09/huffpost-live-fighting-human-trafficking-through-better-data-analysis/.

Achiron, Marilyn (2001). A Timeless Treaty under Attack. UNHCR Refugees Magazine. Volume 2. No. 123. Geneva, Switzerland. p. 6.

Adebiyi, M.A., (2000). Financial Liberalization and Baking Fragility in Nigeria. An empirical investigation. First bank BiB Ann. Rev., 8(17): 59-72.

Adedipe, A., (2005). Building and sustaining corporate performance and growth in the capital market. Nigeria stock exchange annual.

Adegbite, E.O, and S.I. Owualah (2007). "1st National Finance and Banking Conference on Economic Reforms and The Nigerian Financial System", A Communique read at National Conference organised by the Department of Finance, FBA University of Lagos, Nigeria.

Adegbite, I. (1992). "Effective growth and survival of small and medium-scale enterprises in Nigeria in the 1990s and beyond: the role of policy". In Ade, T.O., (1995): "Management of small and medium scale enterprises in Nigeria."Puncmac Publishers, Lagos.

Adejugbe (1987). Report on Issues in small and Medium Enterprises in Nigeria.

Ademuwagun, J. (2011). "Who is Lee Kwan Yew?". Thought Revolution (Tr) Magazine, Volume 1, No. 1.

Aderibigbe, J.O. (1997). "Monetary policy and Financial sector Reform" CBN Bullion, vol. 21, No. 4, pp. 7-13.

Adesanya (1972). Issues on small and medium enterprises in Nigeria.

Afonso Gonçalves da Silva, (2001). "Using the First Principal Component as a Core.

Africa Unity (AU) (1980). Charter on Human Rights.

Africa Unity (AU) (1990). Charter on the Rights and Welfare of the Child.

Africa Unity (AU) Protocol on the Rights of Women in Africa (1981).

African Development Bank (ADB). (2011). The Middle of the Pyramid: The Dynamics of the Middle Class in Africa.

African Journal of Reproductive Health Vol 13 No 2 June (2009). Person in Nepal, Adolescent Girl Survey Questionnaire. New York: Population Council, 2000.

Agarwal, J.P. (1980). "Determinants of Foreign Direct Investment: A Survey", Weltwirtschaftliches Archiv, 117:30-64.

Agbor, J.,Taiwo, O. and J. Smith. (2012). Sub- Saharan Africa's Youth Bulge: A Demographic Dividend or Disaster?

Agence France Presse (2001). 'EU Commission favors temporary asylum for eastern European prostitutes'. March 8. Text available at http://www.pambazuka.org/newsletter.php?category=REFUGEES+AND+FORCED+MIGRATION

Agency for Prohibition of traffick in Persons, Nigeria. Ofuoku AU (2010). Human Trafficking in Nigeria and its Implication for Food Security (In) Int. J. Rural Stud., 17 (1): 1- www.vrionline. org.uk/ijrs (Retrieved April 2011). Poulin R (2004) Globalization and Sex Trade: Trafficking and Agenda: An Address by CBN Governor to the Nigerian Banker Committee.

Aggregate Supply Disturbances", American Economic Review 79: 655-73.

Aggregating Prices in the CPI." Federal Reserve Bank of St. Louis Review(May/June).

Aghion, P., Benerjee A., Piketty, T. (1999). Dualism and Neuroeconomic volatility. Quarterly Journal of Economic 14, 1397 (November).

Agu, C.C. (1988). Nigerian Banking Structure and Performance: The Banking System's Contribution to Economic Development, Onitsha: African-Fep Publishers

Ahmad, N., Laquiller, F., Marianna, P., Pilat, D., Schreyer, P., and Wölfl, A. 2003.

Ahmed, Kamal and Bright, Martin (2002). 'Blunkett to target sex smugglers'. The Observer. February 3. Text Available at www.observer.co.uk/print/0,3858,4348794,00.htmlAIDS: Best Practices Collection.

Ahmed, S. and Mortaza, G. (2005). Inflation and Economic Growth in Bangladesh: 1981-2005. Policy Analysis Unit (PAU) Working Paper 0604.

Aigbokhan, B.E. (1995). Macroeconomics: Theory, Policy and Evidence City: Idelojie publishers.

Aina, T.A. (1996). Globalization and Social Policy in Africa. Codesria Bulletin, 4.

Ajakaiye, O. (2002). "Economic Development in Nigeria: A Review of Experience" CBN Bullion, Vol. 26, No. 1, pp. 47-64.

Ake, C. (1995). The new world order: The view from Africa, in Hans-Henrik and G Sovensen (eds), Whose world order: Uneven globalization and end of coldwar, London, West View Press.

Åkerholm, J. and A Brunila (1995). 'Inflation targeting: the Finnish experience, in

Akinbobola, A. (1999). Globalization and economic development of emergent nation states: A Comparative study of Thailand, Kenya and Taiwan. 1998/99 MILD Journal and Year Book. Lagos.

Akinlami, T. (2011). "African Citizenship Issues". Thought Revolution (Tr) Magazine, Volume 1, No. 1.

Akinrinde, I. (2011). (Unpublished)."Transforming BRICS to BRINCS: thoughts on positioning Nigeria as a major world economic player."Nigerian Economics Students' Association (NESA), Unilag Chapter.

Akpan H. Ekpo (2011). The Economy and Jonathan's Transformation Agenda TELL, Wednesday, 28 September 2011.

Akpan H. Ekpo, Growth without Development in Nigeria, TELL, Thursday, 29, March 2010.

Akpokodje, G. (1998). "Macroeconomic Policies and Private Investment in Nigeria", In: Rekindling Investment for Economic Development in Nigeria by Ben. E. Aigbokhan (ed) Selected Papers for the 1998 Annual Conference, The Nigerian Economic Society.

Alchian, A, and B Klein (1973). 'On a correct measure of inflation', Journal of Money.

Altenburg, T. (2007). Donor Approaches to Supporting Pro-Poor Value Chains. American Economic Review, 57.

Amnesty for Women (2002). 'Geography, Climate, Politics'. Text available at www. femmigration. net/intro_italy.html

Amnesty for Women (2002). 'Italy'. Text available at www.femmigration.net/intro_italy.html

Amnesty International (2002). 'Victims of Human Trafficking Need Help, Not Detention and Deportation'. Text available at http://www.amnesty133.org/ai/action/refugee.html

Amnesty International USA (2000). NSA 6/00 Refugee Concern (Fear of Refoulement).

Amsden, Alice (2001). The rise of the rest: challenges to the west from late-industrializing Economics, oxford university press *an Integrated Supply Chain*: Available from: http:// www.mcb.co.uk.html (Accessed on 20Analysis, Faculty, BIET-MBA Programme, Davangere, Karnataka.

Anao. A.R., (1995). The Accountancy Profession and Nigeria economic development Bank overview. The Nigeria Accountant Official J. ICAN. ISSN 0048-0371. CBN, 2005. Central Bank of Nigeria Annual Report and Accounts. and Development, Paris. and Economic Studies, vol. 15(2):1-26. and M Gilbert, eds., Inflation: the problems it creates and the policies it requires.

Andersson, K, and C Berg (1995). 'The inflation target in Sweden', in Haldane.

Andre S.J. Escrband, A., Molina, C. and Taguas, D. La (1990). Inversun en Espana Econometria con Restricaiones de Equilibro. Barcelona: A Bosh.

Anti-Trafficking Programme (ATP) (2002). Good Practice Studies. Text available at http:// www.antitrafficking.org/best_practice.htm

Anya O. A. (2001). The challenge before the economy. The Post Express, Lagos.

Anyanwu, J.C. (1993). Monetary Economics: Theory, Policy and Institutions, Onitsha: Joanee Educational Publishing Ltd.

Apap, Joanna; Cullen, Peter; and Medved, Felcita (2002). Counteracting Human Trafficking: Protecting the Victims of Trafficking. Centre for European Policy Studies. April 08, 2014. http://www.unodc.org/documents/data-and analysis/glotip/Trafficking_in_ Persons_ 2012_ web.pdf.

Appadurai, A. (2001). Grassroots globalization and the research imagination, in A. Appadurai, (ed), Globalization, Durham: Duke University Press.

Appelel (1981). report on small and medium enterises (SMEs). Applied Economics, Vol. 28: 577–584.

Aremoghu, D. (1993). "Learning About Other's Actions and the investment Accelerator." The Economic Journal, 103 (417):318-328.

Aremu, J.A. (1997). "Foreign Private Investment: Issues, Determinants, Performance and Promotion", CBN Research Seminar Papers, No. 4.

Arestis P., Dennlnades and Luintel K, (2001). Financial Development and Economic Growth: The Role of stock markets, Journal of money, credit and Banking, 33(1) 16-41 in Olofin,

S.O. and Afangidon, U.J. financial structure and Economic Grwoth in Nigeria. A Macro-Economic Approach.

Arestis, P., Dennelmades and Luintel, K. (2001). financial development and Economic growth: the rule of stock market, Journal of money and Banking.

Arlacchi, Pino (2000). UN Office for Drug Control and Crime Prevention. From his statement, as reported in Refugee Reports, Vol. 21. No. 5. Text available at http://www.refugees.org/world/articles/slavery_rr00_5.htm

Aronowitz, Alexis A. (2009). Human Trafficking, Human Misery: *The Global Trade in Human Beings.* Westport: Praeger Publishers.

Aryeetey, Ernest (2003). Resent Development in African financial market: Agenda for further Research: Journal of African Economics 12 supp 2:111-152

Asian Countries, Asian Pacific Development Journal, 8(1), 123-135.

Asian development Bank (2000). The role of central banks in microfinance in Asia and the pacific. ADB.

Associated Press World News (2002). 'Mass Italian police operation snags hundreds of immigrant prostitutes'. Rome, Italy. August 15. Text available at www.protectionproject.org/main1.htm

Athens News Agency (2002). 'Seminar in Athens on Human Trafficking, sex trade'. Athens, Greece. December 12. Text available at www.protectionproject.org/main1.htm

Aucremanne, L., (2000). "The Use of Robust Estimators as Measures of Core Inflation," National Bank of Belgium Working Paper, No. 2. Australia Bulletin, August, pp. 1-6.

Australian Bureau of Statistics (ABS) (2004).'Seasonally adjusted and trend estimates for the consumer price index (CPI)', Australian Economic Indicators, ABS Cat No 1350.0, December, pp 3–8. Available from:http://www.mcb.co.uk.html (Accessed on 13 March 2002).

Ayua (ed.) Proceedings of the National Conference on the Problems of Corruption in Nigeria, 26, 29 March 2001, Abuja.

Ayua, I. (2001). (ed.). Proceedings of the National Conference on the Problems of Corruption in Nigeria, 26, Abuja.

Ayyoub M, Chaudhry, I and Farooq, F. (2011). Does Inflation Affect Economic Growth?

Azam, Jean-Paul, Augustin Fosu, Njuguna S. Ndung'u (1999). "Explaining slow growth in Africa". African Development Review, Volume 14, Issue 2.

B. A. Folorunso and A. G.Abiola, "Long-Run Determi-nants of Inflation in Nigeria (1970-1998)." Nigeria Journal Economics and Social Studies, 42:0-26.

Bacha, E. (1984). "Growth with Limited Supplies of Foreign Exchange: A Reappraisal of the Two Gap Model", The World Bank.

Baddeley, M. (2003). Investment: Theories and Analysis. Basingstoke: Palgrave Macmillan.

Baird, Vanessa (2002). Fear eats the Soul. New Internationalist, Issue 350. October. p. 10.

Baiya, Hzrun and Kithinji, Judith. (2010). Transforming the Dairy Sector: Benefits from the Formalization of the Raw Milk Trade in Kenya. SITE Case Study. Nairobi: SITE Enterprise Promotion.

Baker, Peter (2002). 'Moldovan sold kidney to buy dad's medicine'. Toronto Star Newspapers, Ltd. The Hamilton Spectator. Toronto, Canada. November 9. Text available at www.protectionproject.org/main1.htm

Bakhshi, H., and T. Yates, (1999). "To Trim or Not to Trim, An Application of a Trimmed Mean Inflation Estimator to the United Kingdom," Bank of England, Working Paper Series, No. 97. London, Bank of England.

Bales K. (2002). The social psychology of modern slavery. *Sci Am,* 286, 80-8.

Bales, Kevin (2001). 'Slavery/Economics'. New Internationalist. Issue 337. Text available at www.newint.org/issue337/cheap.htm

Balke N.S. and M. A.Wynne, (1996). "Supply Shocks and the Distribution of Price Changes," Federal Reserve Bank of Dallas, Economic Review, First Quarter, pp.10–18.

Ball L and NG Mankiw (1995).'Relative-price changes as aggregate supply shocks', Quarterly Journal of Economics, 110(1), pp 161–193.

Ball, L and Sheridan, N (2003). "Does Inflation Targeting Matter?" IMF working paper No. 03-129.

Ball, L, and N G Mankiw (1992). 'Relative-price changes as aggregate supply shocks'.

Ball, L, and N G Mankiw (1994). 'Asymmetric price adjustments and economic.

Ball, L., and N.G. Mankiw, (1995). "Relative Price Changes as Aggregate Supply Shocks," Bank of Australia Bulletin, August, pp 1–6.

Baneke, Peter (2002). 'Asylum in Europe'. European Council on Refugees and Exiles. June 9. Text available at www.ecre.org/speeches/bgmsev.shtml

Bank of Canada (1991a). 'Background note on the targets', Bank of Canada Review.

Bank of Canada (1991b). 'Targets for reducing inflation: further operational and Bank of Canada Working Paper 97-9.

Bank of Japan Monetary and Economic Studies, 15(2):1–26.

Bank of New Zealand Discussion Paper G95/5.

Bank of New Zealand Discussion Paper G97/7.

Bank of St. Louis Review(May/June), pp. 143-55.

Bank Target?" Journal of the European Economic Association, Vol. 1, No. 5.

Barro, R.J. (1995). "Inflation and Economic Growth", NBER working Paper Series, No. 5326.

Barrows Jeffre and Finger Reginald (2008). Human Trafficking and the Healthcare Professional. *Southern Medical Journal,* 101, 521-524 10.1097/SMJ.0b013e31816c017d.

Basoglum., Livanou M.,Crnobaric C., Franciskovic T., Suljice., Duric D. and Vranesic M. (2005). Psychiatric and cognitive effects of war in former yugoslavia: association of lack of redress for trauma and posttraumatic stress reactions. JAMA, 294, 580-90.

Bassey, G. E. and E.A. Onwioduokit (2011). "An Analysis of the Threshold.

Bawa, S. and I.S. Abdullahi (2012). "Threshold Effect of Inflation on Economic Growth in Nigeria", To Appear in CBN Journal of Applied Statistics, Vol. 3, No. 1.

BBC News (1998). 'A trade in human misery'. London, UK. December 14. Text available at http://news.bbc.co.uk/1/hi/world/south_asia/234758.stm

BBC News (1998). 'Report says sex trade increasing in Asia'. London, UK. December 14. Text available at http://news.bbc.co.uk/1/hi/world/south_asia/234696.stm

BBC News (1999). 'First lady: End prostitution trade'. London, UK. October 10. Text available at http://news.bbc.co.uk/1/hi/world/europe/471002.stm

BBC News (2000). 'Crackdown on sex slavery'. London, UK. February 23. Text available at http://news.bbc.co.uk/1/hi/world/americas/653648.stm

BBC News (2000). 'International police sacked over Bosnia scandal'. London, UK. November 29. Text available at http://news.bbc.co.uk/1/hi/world/europe/1046929.stm

BBC News (2000). 'U.S. prostitution traffic revealed'. London, UK. Text available at http://news.bbc.co.uk/1/hi/world/americas/698537.stm

BBC News (2001). 'Eastern European women being sold into sex business' London, UK. August 8. Text available at http://news.bbc.co.uk/1/hi/events/newsnight/1484368.stm

BBC News (2001). 'EU 'to protect' sex slaves'. London, UK. March 8. Text available at http://news.bbc.co.uk/1/hi/world/asia-pacific/1037517.stm

BBC News (2001). 'EU cracks down on people-smugglers'. London, UK. January 22. Text available at http://news.bbc.co.uk/1/hi/world/europe/1130243.stm

BBC News (2001). 'FBI urges human traffic co-operation'. London, UK. May 21. Text available at http://news.bbc.co.uk/1/hi/world/europe/1342537.stm

BBC News (2001). 'Helpline for women who head West'. London, UK. March 4. Text available at http://news.bbc.co.uk/1/hi/world/monitoring/media_reports/1201153.stm

BBC News (2001). 'Russian women campaign against trafficking'. London, UK. May 16. Text available at http://news.bbc.co.uk/1/hi/world/europe/1333917.stm

BBC News (2001). 'Sex workers to combat trafficking'. London, UK. March 6. Text available at http://news.bbc.co.uk/1/hi/world/south_asia/1205098.stm

BBC News (2001). 'Sex workers want anti-trafficking measures'. London, UK. March 6. Text available at http://news.bbc.co.uk/1/hi/world/south_asia/1199988.stm

BBC News (2001). 'Sexual slavery rife in Western Europe'. London, UK. July 6. Text available at http://news.bbc.co.uk/1/hi/world/europe/822204.stm

BBC News (2001). 'U.S. decries 'modern-day slavery'. London, UK. July 12. Text available at http://news.bbc.co.uk/1/hi/world/americas/1436329.stm

BBC News (2002). 'Reflex' action against human traffickers'. London, UK. June 25. Text available at http://news.bbc.co.uk/1/hi/uk/2065335.stm

BBC News (2002). 'Rome conference tackles human trafficking'. London, UK. May 15. Text available at http://news.bbc.co.uk/1/hi/world/europe/1988298.stm

BBC News (2002). 'U.S. blacklists trafficking countries'. London, UK. June 5. Text available at http://news.bbc.co.uk/1/hi/world/americas/2027065.stm

BBC News (2002). 'UN warns of human trafficking'. London, UK. February 19. Text available at http://news.bbc.co.uk/1/hi/world/europe/1829855.stm

BBC World Service (2000). 'Frontier sex-trade vigil'. London, UK. November 24. Text available at http://news.bbc.co.uk/1/hi/world/south_asia/1038737.stm

BBC World Service (2000). 'Polish foundation: sex slavery is growing'. London, UK. September 15. Text available at http://news.bbc.co.uk/1/hi/world/europe/926934.stm

BBC World Service (2000). 'U.S. strengthens women's rights'. London, UK. October 6. Text available at http://news.bbc.co.uk/1/hi/world/americas/960523.stm

BBC World Service (2001). 'EU ministers in sex slave talks'. London, UK. February 9. Text available at http://news.bbc.co.uk/1/hi/world/europe/1162449.stm

Bear, Marshall A., Gibbons, Caythy; Haggblade, St even and Ritchie, Nic k. (1992). Facilitator's Guide for Training in Subsector Analysis. GE MINI Tools for Micro Enterprise Programs.

Beaumont, Peter; Morris, Chris and Kaya, Mahmut (2001). 'Bosnia's Corrupt elite grow fat on human cargo smuggled to the West'. The Observer. January 28. Text available at www.observer.co.uk/international/story/06,6903,429995,00.html

Bedont, Barbara (1999). Ending Impunity for Gender Crimes under the International Criminal Court. Democracy and Justice Program, International Centre for Human Rights and Democratic Development. The Brown Journal of World Affairs, Vol. VI, Issue 1: 65-85. Text available at http://www.crlp.org/pub_art_icc.html

Bekaert, G., Harvey, C., Lundblad, C., (2002). Does financial liberalization Spur Growth? Unpublished manuscript, Columbia, University, New York.

Bekaert, G., Harvey, C., Lundblad, C., (2002). Growth volatility and equality market liberalization Duke University working paper.

Ben Uzor, \Tackling Women Tracking through constructive Engagements",

Bennell, P. and Hartl, M. (2011). Investing in the future, Creating opportunities for young rural people. Rome: IFAD.

Bermamke, Bins, S. Thomas Laubach, Adams Posen and Federic S. Misukin (1999). Inflation Targeting : lesson from the intermatiional Experience. Plrlillnceton NJ: Princeton University Press.

Bernanke, B.S., Laubach, T., Mishkin, F.S., and Posen, A.S. (1999). Inflation Targeting.

Bernanke, Ben S., Laubach, Thomas, Mishkin,Frederic, S. and Adam S. Posen, (1999). Inflation Targeting: Lessons from International Experience (Princeton: New Jersey).

Berthiaume, Christiane (1995). 'Do we really care? Refugee Women'. Refugees Magazine, United Nations High Commissioner for Refugees. p. 10-13.

Bertone AM. Sexual Trafficking in Women: International Political Economy and the Politics of Sex. Gender Issues 2000; winter, 18(1):2-22.

Beyer C. (2001). HIV/AIDS in Asia: Accelerating and Disseminating. *The Washington Quarterly,* 24, 211-225.

Beyer C. and Stachowiak J. (2003). Health Consequences of Trafficking ofnWomen and Girls in Southeast Asia. The Brown Journal of World Affairs, 10:105-117.

Beyrer C. (2004). Is trafficking a health issue? *Lancet,* 363, 564.

Billies (1974). Module 1: Gender and food security, Gender in Agriculture Sourcebook, World Bank.

Bishop, Patrick (2001). 'In Paris' . The Daily Telegraph. August 10. Text available at http://www.unhcr.ch/cgi-bin/texis/vtx/home/+uwBmea8vy_eqxwhFq nN0bItFqnDni5AF qnN0bIcFqc BGwhhoqAoDtMzm0wDzmxw/opendoc.htm

Black, Maggie (1993). 'Girls and Girlhood'. New Internationalist. Issue 240. Text available at www.newint.org/issue240/keynote.htm

Blanchard, O.J. and Quah, D. (1989). "The Dynamic Effects of Aggregate Demand.

Blejer, M., Leome, A.M, Rebanal, P. and Schwartz G. (2001). Inflation targeting in the context of IMF adjustment programs, International Monetary Fund Working paper no. 01/31.

Blejer, Mario I., Alfredo M. Leone, Pau Rabanal and Gerd Schwartz, (2002). "Inflation targeting in the Contect of IMF-Supported Adjustment Programs", IMF staff papers. Vol. 49 (30):313-338.

Blejer, Mario I., Leone, Alfredo M., Ruband, Pau and Gerd Schwartz, (1999). "Inflation Targeting in the Context of IMF-Supported Adjustment Program,", In ed. by Loayza, Norman and Raimundo Sato, op. cit., pp. 409–499.

Blinder, Alan S., (1997). "Commentary on Stephen G. Cecchetti (1997)." op. cit., Federal Reserve Bank of St. Louis Review, May/June. pp.157–160.

Blinder, S (1997). 'Commentary', Federal Reserve Bank of St. Louis Review (May/June).

Bloem, Adriaan, Armknecht, Paul, A., and Kimberly D. Zieschang, (2002). "Price Indexes for Inflation Targeting," in ed. By Carson and others (2002) op. cit.,pp.172–198.

Bloom D, (2010). "Prospects for Economic Growth in Nigeria: A Demographic Perspective", Paper presented at the IUSSP Seminar on De mographics and Macroeconomic Performance June 2010.

Bolwig S., Ponte, S., de Toitm A., Riisgaard, L. and N. Halberg. (2010). Integrating Poverty and Environmental Concerns into Value-Ch ain Analysis: A Conceptual Framework. Development Policy Review, 28(2): 173-194.

Bolwig, S., S. Ponte, A. Du Toit, L. Riisgaard and N. Halberg. (2008). Integrating poverty, gender and environmental concerns into value chain analysis: A Conceptual Framework and lessons for Action Research, DIIS, Working Paper no 2008:16.

Book Factor, R. (2001). Supply Chain Management. *The Retail Advisor Fall.* 2001:7.

Boomgard, James J., Davies, Stephen P., Haggblad e, Steven, and Mead, Donald C. (1992). A Subsector Approach to Small Enterprise Pr omotion and Research. World Development, 20(2):199-212. 29.

Botti, Anita (2000). 'The Trade in Human Beings is a Worldwide Scourge'. International Herald Tribune. June 1. Text available at http://www.commondreams.org/view/060100-103.htm

Bouce James and Leonce Ndikumana (2001). "is Africa a net creditor? New Estimate of capital Flight for severally indebted Sub Saharan African countries 1970-1996, Journal of Development Studies, 38(2):27-56.

Bourgeois, Robin and Herrera, Danilo. (2000). Actor-Led Change for Efficient Agri-Food Systems: Handbook of the Participatory Actor-Based CADIAC Approach (Bogor, Indonesia:

CGRPT Centre: Regional Coordination Centre for Research and Development of Coarse Grains, Pulses, Roots and Tuber Crops in the Humid Tropics of Asia and the Pacific).

Bradbury RE. Benin Studies. London: Oxford University Press, 1973.

Braddy, B. (2002). New Avenues for Logistics Cost Reductions: Replacing Inventory With Information. Inbound Logistics Jan. 2002: 14-15.

Bradley, R. (2001). "Finite Sample Effects in the Estimation of Substitution Bias in the Braunstein, Elissa and James Heintz (2006). "Gender Bias and central Bank policy: employment and inflation Reduction" . International review of Applied Economics March, 2008.

Breslau N. 2002. Epidemiologic studies of trauma, posttraumatic stress disorder, and other psychiatric disorders. Can J Psychiatry, 47:923-9.

Bricas, Nicolas, Marie-Cécile Thir ion, and Bernardin Zoungrana. (2009). Bassins de production et de consummation des cultures vivrièrc s cn Afrique de l'Ouest et du Centre. Ougadougou: CILSS.

British Columbia, Department of Economics Discussion Paper Series, No. 95–31.

Brittan, Victoria and Elliot, Larry. (2000). 'Rich Live Longer, Poor Die Younger in a Divided World'. The Guardian. June 29. Text available at http://www.guardian.co.uk/Print/0,3858,4034738,00.html

Brouwer, de, Gordon., (2004). "Discussant" on Heath and others (2004), op. cit.

Brown, L. (2008). Module 1: Gender and Food Security, Gender in Agriculture Sourcebook, World Bank.

Bruinsma G and Bernasco W. (2009). Criminal groups and transnational illegal markets: A Human Sex Trafficking in Nigeria African Journal of Reproductive Health Vol 13(2):47.

Bruinsma G and Bernasco W. (2004). Criminal groups and transnational illegal markets: A more detailed examination on the basis of Social Network Theory. Crime, Law, & Social Change, 41:79-94.

Bruno, M. and W. Easterly (1996). Inflation and Growth: In search of a stable Relationship. Federal Reserve Bank of St. Louis Review, 78(3):139-146.

Bruno, M. and W. Easterly (1998). "Inflation Crisis and Long-Run Growth", Journal of Monetary Economics, 41: 122.

Bruno, Michael and William Easterly. (1998). "Inflation and Growth": in Search of a stable Relationship, "Federal Reserve Bank of St. Louis Review, 78 (3):139-146.

Bruno, Micheal and William R. Easterly (1998). "inflation crises and long-run Growth" Journal of monetary Economics 41:3-26.

Bryan MF and SG Cecchetti (1993).'The consumer price index as a measure of inflation', Federal Reserve Bank of Cleveland Economic Review, 29(4):15–24.

Bryan MF and SG Cecchetti (1994). 'Measuring core inflation', in NG Mankiw (ed), Monetary Policy, University of Chicago Press, Chicago, 195–215.

Bryan MF and SG Cecchetti (1999). 'The monthly measurement of core inflation in Japan', Bank of Japan Monetary and Economic Studies, 17(1):77–101.

Bryan MF, SG Cecchetti and RL Wiggins II (1997). 'Efficient inflation estimation', NBER Working Paper No 6183.

Bryan, M, and C Pike (1991). 'Median Price Changes: an alternative measure approach.

Bryan, M, and S Cecchetti (1993). 'The Consumer Price Index as a measure of inflation'.

Bryan, M, and S Cecchetti (1994). 'Measuring core inflation', Chapter 6 in N G Mankiw.

Bryan, M, S Cecchetti and R Wiggins (1997). 'Efficient Inflation Estimation', NBER.

Bryan, M.F., and Cecchetti, S.G. (1994). "Measuring Core Inflation." In Monetary.

Bryan, Michael F., and Stephen G. Cecchetti, (1996). Inflation and the Distribution of Price.

Bryan, Michael F., and Stephen G. Cecchetti, (1993). The Consumer Price Index.

Bryan, Michael F., and Stephen G. Cecchetti, (1994). "Measuring Core Inflation".

Buch, C.M., Dopke, J., Plerdzioch C.I., (2002). Financial Openness and Business Cycle Volatility, Kid Institute for World Economics Working Paper.

Bullard, J.K. (1995). "The Long-Run Relationship between Inflation and Output in Post-war Economies", Journal of Monetary Economics, (36):477-496.

Burdekin, R.C.K. (2000). "When does Inflation Hurt Economic Growth? Different Nonlinearities for Different Economies", Working Papers in Economics, Claremont Colleges.

Burdett, Kenneth and Judd, Kenneth L.(1983). Equilibrium Price Dispersion, Econometrica, Business and Economic Statistics, (13):253–265.

Burkhalter, H. (2003). Sex Trafficking and the HIV/AIDS Pandemic htt:// physiciansforhumanrights .org/Library/2003-6-25.htm/#(Retrieved April 2011).

Busza J., Castles. and Diarraa. (2004). Trafficking and health. BMJ, 328, 1369-71.

Byerlee, Derek. (2012). Business Models for Promoting the Moderniz ation of Smallholder Value Chains. Washington, DC: Partnership to Cut Hunger and Poverty in Africa.

Byrn, R.T. and G.W. Stone (1992). Economics, U.S.A: Harper Collion.

Cadbury, A., (2002). A Corporate Governance and Chairmanship. Oxford University Press J. Bean, Thear, 2 (3): 112-117.

Caldwell G. (1997). Trafficking of Women from the Former Soviet Union. *The Forced Migration Monitor*, New York: Open Society Institute (Sept.).

Caldwell, Gillian; Galster, Steven; and Steinzor, Nadia (1997). 'Crime and Servitude: An Expose of the Traffic in Women for Prostitution from the Newly Independent States'. Global Survival Network. p. 10. Text available at www.globalsurvival.net/femaletrade/ 9711russia.html

Calstrom, C. and G. Samolysk. (1995). Loan Sales Response to Market Based Capital Requirement. J. Banking Finance, 19: 627-646.

Camara S. (2008). "Determinants Reales financies de la Inversion en Esiana cion unidos Analsis. Teorico Y estimacion Empirica, Bilbao: Xi Jornadas de Economica Critica 2008.

Cambell, Duncan (2000). 'Young girls sold as sex slaves in U.S., CIA says'. The Guardian. April 3. Text available at www.guardian.co.uk/print/0,3858,3981542,00.html

Campbell J., Jones A. S., Dienemann J., Kub J., Schollenberger J., O' Campop., Gielena A. C. and Wynne C. 2002. Intimate partner violence and physical health consequences. *Arch Intern Med,* 162, 1157-63.

Canada, Technical Report No. 89, January.

Canadian Council for Refugees (2000). 'Network on Women Fleeing Gender-Related Persecution Newsletter: Issue 1'. May 29. University of Toronto, Faculty of Law. Text available at www.web.net/~ccr

Canadian Council for Refugees (2000). 'News on Refugee Women Fleeing Gender-Based Persecution'. University of Toronto, Faculty of Law. Text available at http://www.web.net/ ~ccr/newsgend.htm.

Canadian Council for Refugees (2001). 'Network on Women Fleeing Gender-Related Persecution Newsletter', Issue 1. University of Toronto, Faculty of Law. Text available at www.web.net/~ccr

Canadian Council for Refugees (2001). 'Refugee Women Fleeing Gender-based Persecution – International Conference Overview'. University of Toronto, Faculty of Law. Text Available at www.web.net/~ccr/gendpers.html

Caroll, Rory (2001). 'Willing Sex Slaves'. The Guardian. March 1. Text available at www.guardian.co.uk/print/0,3858,4144156,00.html

Carr, M. and Hartl, M. (2010). Lightening the Load: Labour-saving technologies and practices for rural women, IFAD. Carson and others (2002)op. cit.

Carroll, Rory (2002). 'Western Officials 'colluding with people traffickers'. The Guardian. July 23. Text available at www.guardian.co.uk/print/0,3858,4466679,00.html

Carter, M. and Katz, M. (1997). Separate spheres and the conjuga l contract: Understanding the impact of gender-biased development, in Intrahousehold reso urce allocation in developing countries: Models, methods, and policy, The John Hopkins University Press.

Cartwright, Hon Justice Dame Silvia. (2000). Contemporary Gender Issues In The Refugee Context International Association Of Refugee Law Judges. March 10. Text available at http://www.refugee.org.nz/IARLJ3-00Cartwright.html

Carver, T.N. (1903). "Therelation of Abstinence to interest" Quarterly Journal of Economics, 18(1):142-145.

Case of Uganda, (1997). African Economic Research Consor-tium, AERC Research Paper No. 59, Nairobi.

Castles, S. and M. Miller (1998). The Age of Migration: International Population Movements in the Modern World, 2ed. Guildford Press, New York, NY USA.

CBN (1974). "Origin and Development of Inflationary Trends in African Countries: Impact on their Growth", Association of Central Banks seminar held in Addis Ababa, Ethiopia, 5th– 16th August, 1974. CBN Journal of Applied Statistics Vol.3 No.2: 123.

CBN, "CBN Statistical Bulletin," (2008). 50th Anniversary Edi-tion, Lagos, K., Ulrichand K. Frauke, "Data Analys 2nd Edition, Stata Press, Texas.

Cecchetti, S.G. and Debelle, G. (2004).'Has the inflation process changed?', Prepared for the Bank for International Settlements annual conference 'Understanding low inflation and deflation', Brunnen, Switzerland, 18–19 June.

Cecchetti, Stephen G., (1997). "Measuring Short-Run Inflation for Central Bankers," Review Center for Global Development : Publications: Resolving Nigeria's Debt Through a Discounted Buyback.

Center for Gender and Refugee Studies (2001). 'The Case of Ms. O- Update'. Text available at http://www.uchastings.edu/cgrs/campaigns/traffick-pros.htm

Center for Gender and Refugee Studies (2002). 'Asylum for Trafficked Women'. University of California Hastings College of Law. Text available at http://www.uchastings.edu/cgrs/campaigns/traffick-pros.htm

Center for Gender and Refugee Studies (2002). 'Summaries of individual gender asylum cases and decisions, primarily at the affirmative and immigration judge levels'. University of California Hastings College of Law. Text available at http://www.uchastings.edu/cgrs/

Central Bank of Nigeria (2005). Microfinance Policy, Regulator y and Supervisory framework for Nigeria, Abuja, Nigeria. Central Bank Working Paper Series, No. 5. Challenges (Santiago: Central Bank of Chile).

Chambers, Robert. (1994). Part icipatory rural appraisal (PRA): Analysis and Experience. World Development, 22(9): 1253-1268.

Champion Daily. (2007). Nigeria: victims of human trafficking contract AIDS. Available: http://allafrica.com/stories/200708160019.html [Accessed August 17, 2007].

Chan, K. and. R.S. Tsay (1998). Limiting Properties of the Least Squares Changes," National Bureau of Economic Research Working Paper, No. 5793.

Chatfield, C., M. Yar, (1991). "Prediction Intervals for Multiplicative Holt–Winters".

Chaudhry, S., Qamber, Y. and Farooq, F. (2012). Monetary Policy, Inflation and Economic.

Chen, Martha Alter. (1996). Beyond Credit: A Subsector Approach to Promoting Women's Enterprises. Ottawa, Canada : The Aga Khan Foundation.

Chenery, H. and Bruno (1962). "Development Alternatives in an Open Economy: The Case of Israel", Economic Journal, 72.

Chenery, Hollis, B. and Monsue Syrquin (1975). Patterns of Development, 1950-1970 Oxford Economics paper, 50 (563): 72.

Chete, L.M. NI (1998). "Determinants of Foreign Direct Investment in Nigeria", Ser Monograph Series, (7):289-336.

Chhiber, A. and M. Dailami (1990). "Fiscal Policy and Private Investment in Developing Countries. Recent Evidence on Key Selected Issues" PPR Working Paper, No. 559, The World Bank Chicago: University of Chicago Press, 209-237.

Chiguata F., Schnurr J., Jones-Wilson D. and V. Torres (2005). Being 'real' about youth entrepreneurship in Eastern and Southern Africa. ILO SEED Working Paper No. 72, ILO, Geneva.

Chimobi, O.P. (2010). "Inflation and Economic Growth in Nigeria",Journal of Sustainable Development, Vol. 3, No. 2.

Chirinko, Robert (1993). "Business Fixed Investment Spending: Modeling Strategies, Empirical Results and Policy Implications". Journal Economic Literature, 31 (December, 1875-1911).

Cho, Hae Jung and Salas, Angelica (2000). 'Trafficking in Humans Isn't Behind Us Yet'. February 11. Los Angeles Times. Text available at http://www.trafficked-women.org/latimes.html

Cho, Seo-Young. Philipps-University of Marburg. "3P Anti-trafficking Policy Index for (2012). (188 countries) released." Accessed April 15, 2014. http://www.economics-human-trafficking.net/mediapool/99/998280/data/3P_Index_2012_en_press.pdf.

Choudhary, V. (2008). Strengthening Vertic al Linkages Briefing Paper, USAID.

Christiaensen, L., Demery, L. and Kuhl, J. (2010). 'The Evolving Role of Agriculture in Poverty Reduction: An Empirical Perspective', Working Papers UNU-WIDER Working Paper W, World Institute for Development Economic Research UNU-WIDER. 30

Christoffersen, Peter F., and Francis X. Diebold, (1997). "Cointegration and Long-Horizon.

Clamen, Jenn (2002). 'Sex in the city'. The Guardian. July 9. Text available at www.guardian.co.uk/print/0,3858,4457606,00.html Trafficking of Women for Sexual Exploitation: A Gender-Based Coalition to Abolish Slavery and Trafficking (1999). 'Trafficking of women'. Los Angeles, USA. Text available at http://www.trafficked-women.org/berkshire.html

Clarinda R., Gali J. and Gerther, M. (1999). "The Science of Monetary policy: A new Keynesian perspective". Journal of Economic literature 37(4), 1661-1707.

Clark, I. (1999). Globalization and international relations theory, Oxford: Oxford University Press.

Clark, J.M. (1917). "Business Acceleration and Law of Demand. Journal of Political Economy, 1917.

Claus, I (1997). 'A Measure of Underlying Inflation in the United States', Bank of Canada.

Clemen, R.T., (1989). "Combining Forecasts: A Review and an Annotated Bibliography".

Clements, K.W. and H.Y. Izan (1987).'The measurement of inflation: a stochastic approach', Journal of Business and Economics Statistics, 5(3):339–350.

Clements, M.P., and D. F. Hendry, (1993). "On the Limitations of Comparing Mean Square.

Cohen S. and Williamson G. M. (1991). Stress and infectious disease in humans. *Psychol Bull,* 109:5-24.

Coimbra, Carlos, and Pedro Duarte Neves, (1997). "Trend Inflation Indicators,". Coles and Chuta (1984). Gender and Agricultural Value Chains: A review of current knowledge and practice and their policy implications, FAO, ESA Working paper No. 11- 05.

Coleman, J.S. Foundations of Social theory. Cambridge, Massachusetts: Harvard University Press, (1997). 19. Lindenberg S. Homo socio-oeconomicus:

Coles, C. and Mitchell, J. (2010). Gender and Agricultural Value Chains: A review of current knowledge and practice and their policy implications, FAO, ESA Working paper No. 11- 05.Commer. Soc. Sci. 6(2):332-347.

Collins, Judy (2000). 'UN's 'Women 2000': Women are Reshaping the Planet'. Globe Newspaper Company. Text available at http://www.commondreams.org/views/060900-102.htm

Commission of the European Communities (1996). Communication From the Commission to the Council and the European Parliament: On Trafficking in Women for the Purpose of Sexual Exploitation. Document no. COM(96) 567 final. Brussels, Belgium. November, 11.

Commission of the European Communities (1998). Communication From the Commission to the Council and the European Parliament: For Further Actions in the Fight Against Trafficking in Women. Document no. COM(1998) 726 final. Brussels, Belgium. September 12.

Commission of the European Communities (2000). Proposal for a Council Directive on minimum standards on procedures in Member States for granting and withdrawing refugee status. Brussels, Belgium. Document No. /*COM/2000/0578 final – CNS 2000/0238*/

Commission of the European Communities (2002). Proposal for a Council Directive on the short-term residence permit issued to victims of action to facilitate illegal immigration or trafficking in human beings who cooperate with the competent authorities. Brussels, Belgium. Document No. COM(2002) 71 final. February 11.

Commission on Human Rights (2001). Integration of the Human Rights of Women and the Gender Perspective: Traffic in Women and Girls. Document E/CN.4/2001/72. January 31. Commodification of Women and Children. Ca. Women's Stud., 22: 38-43.

Competition." In The Economics of New Goods, ed. T. Bresnahan and R.J. Gordon. Conference Series on Public Policy, 1:19–46.

Connolly, Kate (2001). 'Ceausescu's legacy of abandoned children'. The Guardian. December 12. Text available at www.guardian.co.uk/elsewhere/jounalist/story/0,7792, 617677,00.html

Connolly, Kate (2001). 'West Builds New Iron Curtain'. The Guardian. September 5. Text available at www.guardian.co.uk/elsewhere/journalist/story/0,7792,547257,00.html

Consumer Price Index." Brookings Papers on Economic Activity1, 305-366.

Consumer Price Index." Journal of Official Statistics. 17:369-390.

Convention on the Elimination of all forms of Discrimination against Women (1979). Adopted and opened for signature, ratification and accession by General Assembly resolution 34/ 180 of 18 December 1979. Entry into force 3 September 1981, in accordance with article 27(1). Text available at www.unhchr.ch/html/menu3/b/e1cedaw.htm

Convention on the Rights of the Child (1989). Adopted and opened for signature, ratification and accession by General Assembly resolution 44/25 of 20 November 1989. Entry into force 2 September 1990, in accordance with article 49. Text available at www.unhchr.ch/ htm/menu3/b/k2crc.htm

Cooke, Ben and Kothari, Uma. (2001). Part icipation, The New Tyranny? (London: Zed Books).

Cooke, J.A. (2000). Logistics ropes in Inventory. *Management and Distribution Report*. July.

Cooley, Thomas F. and Mark Dwyer, (1998). "Business Cycle Analysis without Much Theory.

Corbin B. Lyday, (2001). The Shadow Market in Human Beings: An Anti-Corruption Perspective, Paper Presented at Transparency International's 10th International Anti-Corruption Conference, Prague, Czech Republic, October 7:11.

Corbin B. Lyday, op. cit.. Benjamin Ike, \The Corrupt Practices and other Related Offences Act", (2000). New Nigerian, 15 September 2000.

Coyle, J.J. Bardi, E.J. and Langley, C.J. (1996). The Management of Business Logistics: 6th.

Crainic, G.T. (2002).A survey of optimization models for Long-Haul freight Transportation. Credit and Banking, 5(1): 173-91.

Crawford M. and Kaufman M. R. (2008). Sex trafficking in Nepal: survivor characteristics and long-term outcomes. Violence Against Women, 14: 905-16.

Crawley, Heather (2000). 'Gender and displacement'. Forced Migration Review. Issue 9. December. Text available at www.fmreview.org

Cullen-Dupont, Kathryn. Global Issues: Human Trafficking. New York: Infobase Publishing,

Cunningham, John (2002). 'Asylum seekers drawn into child prostitution, warns expert'. The Guardian. January 30.Text available at www.society.guardian.co.uk/print/0,3858,4345180,00.html

Curtain, R. (2001). Youth and employment: a pub lic policy perspective. Development Bulletin, 55: 7-11.

Cutler, J. (2001).'Core inflation in the UK', External MPC Unit Discussion Paper No 3, Bank of England.

Cutler, Joanne, (2001). "A New Measure of Core Inflation in the U.K.," External MPC Unit.

Da Silva, Carlos A. and da Souza Filho, Hildo M. (2007). Guidelines for Rapid Appraislas of Agri-food Chain Performance in Deve loping Countries. Rome: FAO.

Daily Champion, 5 April 2001, p. 21.

Daily Times, 14 June 2001, p. 4.

Daily Times, 8 October 2001, p. 28. Yomi Odunuga, \Ex-police Chief, 50 others arrested for human trafficking"

Dandona R, Dandona L., Kumar G. A., Gutierrez J. P., Mcpherson S . & Bertozzis. M. (2005). HIV testing among female sex workers in Andhra Pradesh, India. AIDS, 19, 2033-6.

Davis K. 2006. Human trafficking and modern day slavery in Ohio. Washington DC: Polaris Project. December 18: Human Rights of Migrant Workers (2002). 'Update on the development of a common policy of migration', Memo 02/206. Brussels, Belgium. October 7. Text available at http://www.december18.net/instrumentsregionEUupdate Oct2002.pdf

Davis, P. and Drum, F. (1996). Council of Logistics Management Annual Conference, Oak.

Deen, Talif. (2000). 'Trafficking in Human Beings Reprehensible, Says UN's Kofi Annan'. IPS. Text available at www.commondreams.org/headlines/121400-02.htm *Development* 41 (2013).

DeJanvry, A. and Sadoulet, E. (2009). Agricultural growth and poverty reduction: additional evidence. Oxford: Oxford University Press.

Dejuan, O. (2005). "Paths of Accumulation and Growth: Towards a Keynesian Long-period Theory of Output. "Review of Political Economy, 17 (2): 231-252.

DFID (2008). Making Value Chains Work Better for the Poor: A Toolbook for Practitioners of Value Chain Analysis. London: DFID.

DFID and SDC. (2008). The Operational Guide for the Making markets Work for the Poor (M4P) Approach. London and Berne: DFID and SDC.

DFID Policy Paper Dec. (2005). Growth and Poverty Reduction: The Role of Agriculture Published by the Department for International Development.

Diao, Xinshen, Thurlow, James, Benin, Samuel and Fan, Shenggen. (2012). Strategies and Priorities for African Agriculture: Economy wide Perspectives from Country Studies.

Dickey, D.A. and Fuller, W.A. (1981). Likelihood Ratio Statistics for Autoregressive Time.

Diewert WE (1995).'On the stochastic approach to index numbers', University of British Columbia Discussion Paper No 95-31.

Diewert WE (1997).'Commentary for "Alternative strategies for aggregating prices in the Consumer Price Index", Federal Reserve Bank of St. Louis Review, 79(3):126–138.

Diewert, W. Erwin, (1995). "On the Stochastic Approach to Index Numbers,".

Diewert, W. Erwin, (2004). "Basic Index Number Theory," Chapter 15, pp. 263-287 in ILO.

Diewert, W. Erwin, (2004). "Elementary Indices," Chapter 20, pp. 355–370 in ILO and others

Diewert, W.E. (1976). "Exact and Superlative Index Numbers." Journal of Econometrics.

Diewert, W.E. (1983). "The Theory of the Output Price Index and the Measurement of Discussion Paper, No. 3 (March).

Dixon R and GC Lim (2004).'Underlying inflation in Australia: are the existing measures satisfactory?', Economic Record, 80(251), pp 373–386. Do We Stand?" Journal of Economic Literature 41: 159-201.

Doezema, Jo (2000). Editorial. Network of Sex Work Projects. The Guardian. June 2. Text available at www.guardian.co.uk/letters/story/0,3604,327218,00.html

Doguwa, S. I. (2012). Inflation and Economic Growth in Nigeria: Detecting the Threshold.

Dorrance, G. S. (1963). "The Effect of Inflation on Economic Development" IMF Staff Papers, 10: 1-47.

Douglason, G.U and Gbosi, A (2006). The Dynamics of productivity and unemployment Nexus: Implications for employment generation in Nigeria NES 2006". Annual conference, Ibadan, Nigeria.

Dow, J (1994). 'Measuring inflation using multiple price indexes', mimeo, Department of Drugs," Journal of Political Economy, 108(4):833–850.

Drukker, D, Gomis-Porqueras, P. and Hernandez-Verme, P. (2005). "Threshold Effects in the Relationship between Inflation and Growth: A New Panel-Data approach" http://www.uh.edu/~cmurray/TCE/papers/Drukker.pdf

Dunning, J.H. (1973). "The Determinants of International Production", Oxford Economic Papers, 25:289-336.

Dunning, J.H. (1981). "Explaining the International Direct Investment Position of Countries: Towards a Dynamic or Developmental Approach" *Wiltwirt-schaftliches Archiv* 117:30-64.

Dynamics and Control, 12 (2): 231-254. doi:10.1016/0165-1889(88)90041-3.

E. J. Hamilton Flation (1977). "The Role of War in Modern In Journal of Economic History", 37(1):13-19.

Easterly W., Islam, R.J., Stiglitz, J.E., (2001). "Suaken and stirred" Explaining growth volatility In: Pleskovik B., Stern, N. (Eds). Annual World Bank Conference on Developing Economics.

Easterly, William (2002). "An Identity Crisis? Testing IMF Financial Programming". Center for Global Development. Working Paper No. 2, August.

Ebirim, Innocent. \Human Tracking: Economic Implications", New Nigerian,

Ebony Consulting International (ECI). (2001). The Kenyan Dairy Subsector: A Study for DfID/Kenya. Nairobi: DfID.

Eckstein, O (1981). Core Inflation, Prentice-Hall, (Englewood Cliffs, N.J.).

Economic Community of West African States (ECOWAS) (2001). Declaration of Plan of Action against Trafficking in Person.

Economic Trends, No. 577 (December), Reprinted in Carson and others (2002) op. cit.

Economics and Statistics, 52:169-210.

Economics Discussion Paper 95-31, University of British Columbia.

Economics, 32:485-512.

Edo State House Assembly, A bill on sexualtrafficking. Benin City: Edo State, Government, (2002). Caldwell G, Galster S, Kanics J and Steinzor N. Capitalizing on Transition Economies: the role of the Russian Mafiya in Trafficking Women for Forced Prostitution. Transnational Organized Crime 1997, 3(4): 42–73.

Edozien, E.C. (1968). "Linkages: Direct Foreign Investment and Nigeria's Economic Development", Nigerian Journal of Economic and Social Studies 10(2):191-203.

Effects of Inflation on Economic Growth in Nigeria", Waifem Review, 8:2.

Egeln, J. (2000). Die volkswirtschaftliche Bedeutung junger Unter nehmen. In Buttler, G.H., et al. (eds.) Existenzgründung. Rahmenbedingungen und Strategien , Heidelberg : Physica-Verlag.

Eggen, Dan (2000). 'Proposal Broadens Asylum'. The Washington Post. December 8. Text available at http://www.uchastings.edu/cgrs/documents/media/wp_12-00.htm

Ehlers A., Maercker A. and Boos A. (2000). Posttraumatic stress disorder following political imprisonment: the role of mental defeat, alienation, and perceived permanent change. *J Abnorm Psychol,* 109:45-55.

Ekpo, A.H. (1997). "Foreign Direct Investment in Nigeria: Evidence from Time Series Data", CBN Economic and Financial Review, Vol. 35, No. 1.

Elbehri, Azia and Lee, Mari a. (2011). The Role of Women Producer Organizations in Agricultural Value Chains: Practical Lessons from Africa and India. Rome: FAO.

El-Bushra, Judy (2000). 'Gender and forced migration: editorial'. Forced Migration Review. Issue 9. December. Text available at www.fmreview.org

El-Namaki, M. S.S.(1989). "Management Training for Small and Medium Industries: A Cross-country analysis of characteristics, constraints and catalysts" In Proceedings of the Second Tokyo Conference on Management Development of Small and Medium-sized Enterprises in Asia, The Foundation for Asian Management Development, Tokyo.

Engel, Matthew (2002). 'U.S. soft-pedals on allies' human rights record'. The Guardian. March 5. Text available at www.guardian.co.uk/print/0,3858,4367801,00.html

Engle, R. (1982). "Autoregressive Conditional Heteroscedasticity with estimates of the variance of United Kingdom Inflation", Econometrica, 50 (1): 987-1007.

Epstein, Gerald (2007a). Central Bank as Agents of Economic Development", in Ha-Joon changed. Institutional change the economic development, (2007), United Nations, University and Anthem Press.

Epstein, Gerald, (2003). "Alternatives to Inflation Targeting Monetary Policy for stable and Egalitarian Growth: A Brief Research Summary. PERI Working paper. No. 62, www.umass.edu/peri>

Equation". Mimeo, Department of Ecpnomics, University of Hawai'1 at Monoa, Honolulu. Available at www.hawaii.edu/~malla

Espasa, A. and S. Enra, E. (1993). "Consideraciones sobre ia function de inversun en Espana working paper, 1993.

Espinoza, R., Leon, H. and Prasad, A. (2010). "Estimating the Inflation-Growth Nexus – A Smooth Transition Model" IMF Working Paper Estimator of a Continuous Threshold Autoregressive Model, Biometrika, 85(2): 413-426.

Euro Area," Applied Economic Letters, (9): 353–357.

European Commission (2001). Discussion paper on Prevention of Trafficking in Human Beings. May 17/18. http://europa.eu.int/comm/dgs/justice_home/index_en.htm

European Commission Directorate General for Justice and Home Affairs (2002). Tampere: Kick Start to the EU's policy for justice and home affairs. As presented at the European Conference on Preventing and Combating Trafficking in Human Beings. Brussels, Belgium. September 18-20.

European Commission Directorate General for Justice and Home Affairs (2002). Trafficking in human beings: The European Response. Background Document. As presented at the European Conference on Preventing and Combating Trafficking in Human Beings. September 18-20. Brussels, Belgium.

European Council (1999). Presidency Conclusions: Tampere European Council. October 16. Brussels, Belgium. Document no. SN (1999) 800.

European Council on Refugees and Exiles (ECRE) (1997). Position Paper – Position on Temporary Protection in the Context of the Need for a Supplementary Refugee Definition. March. Text available at http://www.ecre.org/statements/ecreuae.shtml

European Council on Refugees and Exiles (ECRE) (2000). ECRE's Comments on the French Presidency proposals for a Council Directive defining and Framework Decision on preventing, the facilitation of unauthorized entry, movement and residence. November 7. Text available at http://www.ecre.org/positions/tp.shtml

European Council on Refugees and Exiles (ECRE) (2000). Position on the Interpretation of Article 1 of the Refugee Convention. September. Text available at http://ecre.org/positions/csrinter.shtml.

European Council on Refugees and Exiles (ECRE) (2000). Position Paper - Women. Text available at http://www.ecre.org/positions/women.shtml

European Council on Refugees and Exiles (ECRE) (2002). The Promise of Protection: Progress toward a European Asylum Policy since the Tampere Summit.

European Forum for Urban Security. "Putting Rantslev into Practice." Accessed April 22, 2014. "http://efus.eu/files/2013/03/Invitation-to-Putting-Rantslev-into-Practice-16-18-April-2013-Amsterdam-the-Netherlands.pdf.

European Journal of Accounting Auditing and Finance Research 1(4):139-150, December 2013. Published by European Centre for Research Training and Development UK (www.ea-journals.org).

European Union (2001). Meeting of JHA Ministers of the Member States and the candidate States: 12 Commitments in the Fight against Trafficking in Human Beings. Brussels, Belgium. September 28. Document no. SN 4026/1/01. Rev. 1.

European Union (2001). Trafficking in Women. The misery behind the fantasy: from Poverty to Sex Slavery. A Comprehensive European Strategy. Brussels, Belgium. March 8. Text available at http:/europa.eu.int/comm/employment_social/equ_opp/index_en.htm

European Union Press Release (2002). Combating illegal immigration and trafficking in human beings: Commission's proposal for a residence permit for victims who cooperate with the authorities reminds the Member States that the phenomenon cannot be tackled at national level alone. Brussels, Belgium. February 12. Text available at www.europa.eu.int/rapid/start/cgi/guesten.ksh?p_action.gettxt=gt&doc=IP/02/224|0|AGED&lg=EN&display

European Union Press Release (2002). EU urges higher priority for fight against trafficking in women and children. Brussels, Belgium. Text available at www.europa.eu.int/rapid/start/cgi/guesten.ksh?p_action.gettxt=gt&doc=IP/01/325|0|AGED&lg=EN&display

European Women's Lobby (2000). Trafficking in women: EWL press release. June 16. Text available at http://www.womenlobby.org/

European Women's Lobby (2000). Trafficking in women: the EWL calls on governments and NGOs to support the implementation of the United Nations 1949 Convention on the trafficking in Human Beings. May 17. Text available at http://www.womenlobby.org/document.asp?DocID=13

European Women's Lobby (2001). Migration, trafficking and social development, what is at stake for women. Brussels, Belgium. Text available at www.womenlobby.org/document.asp? DocID=315

European Women's Lobby (2001). The links between trade and human rights. Migration, trafficking and social development, what is at stake for women? June 6. Text available at http://www.womenlobby.org/document.asp?DocID=315

Evidence from Four South Asian Countries", Asia-Pacific Development Journal, 8(1).

Expenditure in Nigeria: A Historical Survey," An Unpublished M.Sc. Thesis.

Faber SM and AM Fischer (2000).'Core Inflation in Switzerland', Swiss National Bank Quarterly Bulletin, December, pp 66–75.

Fabre, P. (1993). Note de méthodologie générale sur l'analyse de filière. Rome: FAO.

Falemi-Ozcan, S., Sorensen B.E., Yosna, O., (2003). Risk-sharing and industrial specialization regional and International audience. American Economic Review as, 903-918 (June).

FAO, IFAD, MIJARC. (2012). Summary of the Findings of the FAO/IFAD/MIJARC Joint Project on 'Facilitating Access of Rural Youth to Agricultural Activities'. Background Paper, Youth Session of the 2012 Farmer' Forum.

FAO. (2010). Gender dimensions of agricultural and rural employment: Differentiated pathways out of poverty, Status, trends and gaps, Path ways Out of Poverty Program. Rome: FAO.

FAO. (2011). Guidance on how to address rural employment and decent work concerns in FAO country activities. Guidance Material #1. Rome: FAO.

Faria, J. R. and Carneiro, F.G. (2001). "Does High Inflation Affect Growth in the Long and Short Run?" Journal of Applied Economics, 4(1).

Farley M. and Barkan H. (1998). Prostitution, violence, and posttraumatic stress disorder. Women Health, 27:37-49.

Fase, M, and C. Folkertsma (1996). 'Measuring inflation: an attempt to operationalize Carl.

Fatas, A. (2002). The effect of Business cycles on Growth in Loayza, N. Soto, R.C. (ed) Economic Growth Sources, Treadsanol cycles, Central Bank of Chille, Santiago.

Fawcett, P. Mcleish, R. and Ogden, I. (1992). Logistics Management, Singapore.

Fazzari, (1987). S.M. and Mott, T.L. "The Investment Theories of kalacki and Keynes: An Empirical study of firm Data, (1970-82). Journal of Post Keynesian Economics Winter 1986-87, 9(2): 171-187.

Fazzari, G., R.G. Hubbard and B.C. Petersen (1988a). "Financing Constraints and Corporate Investment", Brooking Papers on Economic Activity, Vol. 1 No. 2.

Fazzari, G., R.G. Hubbard and B.C. Petersen (1988b). "Investment, Financing Decisions and Tax Policy", American Economic Review, 82(3).

Federal Government of Nigeria, The (1999). Nigerian Census. Lagos: Government of Nigeria Printing Press, 1999.

Federal Republic of Nigeria: National Rolling Plan, 1991-93.

Federal Reserve Bank of Cleveland Economic Review (4):15-24.

Ferrand, D., Gibson, A., and Scott, H. (2004). Making Markets Work for the Poor: An Objective and An Approach for Governments and Development Agencies. Woodmead: Commark Trust. 31.

Filho, Alfredo Saad (2004). "Pro-poor Monetary and Anti-Inflation Policies Developing Alternatives to the New Monetary Policy Consensus". Mimeo. SOAD, University of London.

Final Report of the Expert Group Entrepreneurship in Higher Education, Especially within Non-business studies. Brussels: European Commission, March 2008.

Finland Discussion Paper 24/94.

Fischer, S. and Modigliani, F. (1978). "Towards an Understanding of the Real Effects and Costs of Inflation", NBER Working Papers, No. 0303.

Fischer, S. (1993). "The Role of Macroeconomic Factors in Economic Growth" Journal of Monetary Economics, 32: 485-512.

Fisher I. (1930). The theory of interest, New York. Macmillian 1930.

Fisher S. (1993).The Role of Macroeconomic Factors in Growth, Journal of Monetary.

Fisher, I. (1911). The Purchasing Power of Money, 2nd edition (1920), Macmillan (New York).

Fisher, I. (1922). The Making of Index Numbers, Houghton Mifflen (Boston).

Fisher, S. (1993). "The Role of Macroeconomic Factors in Growth", Journal of Monetary Economics Vol. 32 (December) pp.458-512.

Flemming, J. (1976). Inflation, Oxford University Press (London) Fluctuations', The Economic Journal, 104:247-61.

Folkertsma, C. K., and K. S.E.M. Hubrich, (2000). "Performance of Core Inflation for Inflation Targeting Regimes and the Role of IMF Data Initiatives," pp. 308-339 in Forecast Errors," Journal of Forecasting, (12): 617–637.

Fonck K., Kaulr R., Keli F., Bwayo J. J., Ngugi E. N., Mosess. and Temmerman M. (2001). Sexually transmitted infections and vaginal douching in a population of female sex workers in Nairobi, Kenya. Sex Transm Infect, 77:271-5.

Forecasting," Federal Reserve Bank of Philadelphia, Working Paper No. 97–14.

Foreign Investment Advisory Service (FIAS). (2007). Moving Toward Competitiveness: A Value Chain Approach. Washington, DC: The World Bank.

Foresight Africa: Top Priorities for the Continent in (2012). Africa Growth Initiative at Brookings, pp.9-11.

Forsyth, F.G. and Fowler, R.F. "The Theory and Practice of Chain Price Index Numbers" in Journal of the Royal Statistical Society A, 144(2): 224-247.

Fourth World Conference on Women (1995). Platform for Action: Strategic Objective E.5 (147 (h)) Beijing, China.

Fowler, B. and M. Brand. (2011). Pathways out of Poverty: Applying key principles of the value chain approach to reach the very poor, USAID, Microreport, 173.

Freeman, D.G., (1998). "Do Core Inflation Measures Help Forecast Inflation?" Economics.

Frenkel, Reberto and Rapetti, Martin (2008). "Five years of competitive and stable real exchange rate in Argentina, 2002-2004" International Review of Applied Economics.

Frenkel, Roberto and Martin Rapetti (2008). "Monetary and Exchange rate Policies in Argentina after the Convertibility Regime Collapse", Intentional Review of Applied Economic. March.

Friedman, M (1963). Inflation. Causes and consequences, Asia Publishing House (New from:http://www.ascet.com/html (Accessed on 26 September 2003)

Friedman, M., (1973). Money and Economic Development, Toronto: Lexington Books.

Fukuyama , F. (1992). *The end of History and the Last Man*, London: Hamish Hamilton.

Fund (IMF) Staff Working Paper WP/00/58, Washington D.C., IMF.

G. Itua, "Structural Determinants of Inflation in Nigeria (1981-1990)." An Unpublished M.Sc. Thesis, Department of Economics, Ahmadu Bello University, Zaria, 2000.

Gartner, C, and G Wehinger (1998). 'Core inflation in selected European Union countries',

Gaspar, Vítor, (2002). "Eurostat's HICP and the European Central Banks' Definition of Price.

Gbosi, A.N. (2006). Modern Labour Economics and Policy Analysis . Abakaliki, Pack Publishers.

Gedda, George (2002). 'Terror concerns cause decline in refugee admissions'. Associated Press. September 3. Text available at http://www2.dailycamera.com/bdc/nation_world_news/article/0,1713,BDC_2420_1364733,00.html

Gentleman, Amelia (2001). 'Russians' hopes dashed on London streets'. The Guardian. June 13. Text available at www.guardian.co.uk/elsewhere/journalist/story/0,7792,506263,00.html

Gereffi G., Humphrey, J., and T. Sturgeon. (2005). The Governance of Global Value Chains. Review of International Political Economy, 12(1): 78-104.

Gereffi, G. (1995). "Global Production Systems and Third World Development." In: B. Stallings (ed.), Global Change, Regional Response: The New International Context of Development, New York, Cambridge University Press, pp.100-142.

Gereffi, G., Lee J., and M. Christian. (2009). US-Based Food and Agricultural Value Chains and Their Relevance to Healthy Diets. Journal of Hunger & Environmental Nutrition, 4(3): 357-374.

Gesher, Ari. Palantir. "HuffPost Live: fighting human trafficking through better data analysis."

Gesterberg (1959). "Small Firms' Training and Competitiveness: Building Upon the Small Business as a Learning Organisation", In: International Small Business Journal 15(3), April-June 1997.

Gibb, A. (1991). "Defining Success in Entrepreneurship Development Programmes: A Guide to Model Approach", ILO, Geneval, 1991.

Gibb, A. (1997). "Small Firms' Training and Competitiveness: Building Upon the Small Business as a Learning Organisation", in International Small Business Journal, 15, No. 3, April-June 1997.

Gibb, A., Hannon, P. (2006). Towards the Entrepreneurial University? Durh am, Birmingham: National Council for Graduate Entrepreneurship.

Gilbert Gallopin (1994). Impoverishment and Sustainable Development: A System Approach. A report of the International Institute for Sustainable Development, IISD, Manitoba, Canada

Gillan, Audrey and Ward, Lucy (2000). 'Prostitutes imported into slavery'. May 30. Text available at www.guardian.co.uk/print/0,3858,4023464,00.html

Glickstein Barbara (2008). The Tragedy of Human Trafficking. AJN The American Journal of Nursing, 108, 11 10.1097/01.NAJ.0000339133.08340.e9.

Godfrey, M.. (2006). Youth Employment in Sub-Saharan Africa, An assessment of existing Interventions, World Bank, Workshop, February 2005. Washington, DC: World Bank.

Goldberg, Ray A. (1968). Agribusiness Coordination: A Systems Approach to the Wheat, Soybean and Florida Orange Economies (Boston: Harvard University Graduate School of Business Administration).

Goletti, Francesco. (2005). Agricultural Commercialization, Value Chains and Poverty Reduction. Making Markets World Better for the Poor, Discussion Paper No.7 (Hanoi: Asian Development Bank).

Goliath JP (2008). Human Trafficking: South Africa Perspective. A paper delivered at the International Association of Women Judges 9th Biennial International Conference, Panama.

Goodhart, C (1994). 'What should central banks do? What should be their macroeconomic.

Gordon, R (1985). 'Understanding inflation in the 1980s', Brookings Papers on Economic.

Gordon, R.J. (2005). "Apparel Prices 1914-93 and the Hulten/Brueghel Paradox.".

Gosh, A. and Phillips, S. (1998). "Warning: Inflation May Be harmful to your Growth", IMF Staff Papers, 45(4).

Granger, C.W.J. (1986). "Developments in the Study of Co-integrated Variables", Oxford Bulletin of Economics and Statistics, 48:213-228.

Granger. C.W.J.. (1969). An Investigating Causal Relations by Econometric Models and Cross Spectral Methods. Econometrical, 37:428-438.

Granier, Karine (2002). The European Union's Big Bang. International Herald Tribune. December 14-15. p. 5.

Grant, William. (2011). Formalizing the Warm Milk Industry in Kenya. Nairobi: Micro-Links. http://microlinks.kdid.org/breakfast.

Green, Anke (2003). "Credit Guarantee schemes for Small Scale Enterprises: An Effective Instrument to Promote Private Sector- led Growth" UNDO Working Paper, SME Technical Working Paper Series.

Greene, J. and D. Villannueva (1991). "Private Investment in Developing Countries: An Empirical Analysis", IMF Staff Papers Vol. 38, No. 2 (March) pp.33-58.

Greenlees, John, and Bert Balk, (2004). "Errors and Bias," Chapter 11, pp. 207–214 in ILO

Greenspan. A. (2004). A Remarkl at the Conference on Bank Structure and Com pennon. Sponsored by the Federal Reserve Bank of Chicago Illinois via Satellite.

Greenwood, J. and R. Jovanovic (1990). "Financial Development, Growth and Distribution of Income", Journal of Political Economy, 98:1076-107.

Greunning, H.V. and S.B. Bratanovic, (2003). A Analysis and Managing banking Risk:, A Framework for Assessing Corporate Governance and Financial Risk. 2nd edn. International Bank for Reconstruction and Development/the World Bank 1818 H Street N.W Washing DC 20433.

Griffin, Jo. The Guardian. "Child sex tourism warning for fans attending World Cup in Brazil."
Gruppo Abele, (2001). Annuario Sociale 2001. Giangiacomo Feltrinelli Editore, Milan, Italy.
Gushulak B. D. and Macpherson D. W. (2000). Health issues associated with the smuggling and trafficking of migrants. J Immigr Health, 2:67-78.
Hadjmichael, M., D. Ghura, M. Muhleisen, R. Nord and E. Ucer (1995). Growth, Savings and Investment, 1986-93. New York: International Monetary Fund.
Hadley E. (2001). Prostitution: solutions for a global problem. Gender Policy Review; 4(2).
Haggblade, Steven and Gamser, Matthew. (1991). A Field Manual for Subsector Practitioners. GEMINI tools for Microe nterprise Programs. Wash ington, DC: Development Alternatives Inc.
Haggblade, Steven. (2007). Subs ector Supply Chains: Operationa l Diagnostics for a Complex Rural Economy. Chapter 15 in Haggblade, Haze ll and Reardon, editors Transforming the Rural Nonfarm Economy (Baltimore, MD: Johns Hopkins University Press).
Haines, Rodger (1996). Gender-Based Persecution: New Zealand Jurisprudence. Presented at the Symposium on Gender-Based Persecution organised by the Office of the United Nations High Commissioner for Refugees. New Zealand Refugee Status Appeals Authority. February 22 and 23.
Haines, Rodger (1998). Interim Report on Membership of a Particular Social Group Prepared for the Inter-Conference Working Party on Membership of a Particular Social Group. Refugee Status Appeals Authority New Zealand. October. Text available at http://www.uchastings.edu/cgrs/documents/media/ir_5-02.doc
Haines, Rodger (2001). Gender-Related Persecution. Prepared for the UNHCR San Remo Expert Roundtable. New Zealand Refugee Status Appeals Authority. August 10. Text no. Confer/UNHCR003
Haldane, A, and C Salmon (1995). 'Three issues on inflation targets', in Haldane, 100.
Haldane, A, ed. (1995). Targeting inflation, Bank of England (London).
Hampel, F.R., E.M. Ronchetti, P.J. Rousseeuw and W.A. Stahel (1986). Robust Statistics: The approach Based on Influence Functions, Wiley, New York.
Hamre C. (2008). Selecting Pro-Poor Value Chains : A Useable Framework. Dissertation. London
Harding, Gareth (2001). 'EU outlines asylum plan for victims of sex slavery' The Independent (U.K.). March 9. Text available at http://www.uchastings.edu/cgrs/documents/media/uk_indt_3-01.htm
Hardwick, Nick (2002). 'Stand up and celebrate asylum'. The Observer. June 26. Text available at www.observer.co.uk/comment/story/0,6903,737560,00.html
Harper, M. (1983). "Selection and Training for Entrepreneurship Development" ILO, Geneva.
Harrison, Ann. (1996). "Openness and Growth: a Time - Series, Cross-Country Analysis for Developing Countries". Journal of Development Economics, 48:419-447.
Harter, H.L., (1974 and 1975). "The Method of Least Squares and Some Alternatives,".
Hartwich, F. and Kormawa, P. (2009). Value chain diagnosis: Building blocks for a holistic and rapid analytical tool, UNIDO, Italy 32.
Hausman, J. (1997). "Valuation of New Goods Under Perfect and Imperfect.
Hawkes, C. and M.T. Ruel. (2011). Value Chains for Nutrition. 2020 Conference Paper 4, prepared for the IFPRI 2020 Internationa l Conference "Leveraging Agriculture for Improving Nutrition and Health", February 10-12, New Delhi, India.
Hayashi, F. (1982). "Tobin's Marginal Q, and Average Q: A Neoclassical Integration", Econometrica, 50:1.
Heath A, ID Roberts and TJ Bulman (2004).'Inflation in Australia: measurement and modelling', in C Kent and S Guttmann (eds), The future of inflation targeting, Proceedings of a Conference, Reserve Bank of Australia, Sydney, pp 167–207.
Heath, Alexandra, Roberts,Ivan, and Tim Bulman, (2004)."Inflation in Australia.

Hedges, Chris (2000). 'Asylum Policy Is Piecemeal in Cases of Gender Abuse'. New York Times. March 12.Text available at http://www.uchastings.edu/cgrs/documents/media/nyt_3-12.htm

Hefti-Misa, Anny (2002). 'Empowerment Strategies of Black, Migrant and Refugee Women'. Philippine Women's Network in Europe. Text available at http://www.univie.ac.at/Voelkerkunde/apsis/aufi/statwom.htm#traffick

Hemmati, Minu. (2002). Multi-Stakeholder Pro cesses for Governance and Sustainability-Beyond Deadlock and Conflict (London: Earthscan).

Henson, S., S. Jaffee, J. Cranfield, J. Blandon and P. Siegel. (2008). Linking African smallholders to high-value markets: Practitioner perspectives on benefits, constraints and intervention, World Bank, Policy Research Working Paper 4573.

Hicks, J.R. (1940). "The Valuation of the Social Income." Economica 7: 105-124.

Hill, Robert J., (2004). "Inflation Measurement for Central Bankers, in The Future of Inflation

Hirschi T. (1969). Causes of Delinquency. Berkeley, California: University of California Press.

Hnatkouska, V., Loayza, N., (2005). Volatility and growth in Alzenman, J., Pinto, B. (eds), Managing Economic volatility and crises: A practitioner's guide. Cambridge University Press.

Hoff, Suzanne. Council of Europe. "The role of NGOs in combating human trafficking and supporting (presumed) trafficked persons." Accessed April 23, 2014. http://www.coe.int/t/dghl/cooperation/economiccrime/trafficking/Projects/THB percent20Azerbaijan/REPORT_HOFF.pdf.

Hogan S, M Johnson and T Laflèche (2001).'Core inflation', Bank of Canada Technical Report No 89.

Hogan, Seamus, Johnson, Marianne,and Thérèse Laflèche, (2001). "Core Inflation,".

Hogg, R., (1967). "Some Observations on Robust Estimation," Journal of the American.

Holliday, Lucy (1994). 'Prostitution'. New Internationalist. Issue 260. Text available at www.newint.org/issue260/update.htm

Holmes, Paul and Berta, Krizstina (2002). Comparative Matrix on Legislation and Best Practices in Preventing and Combating trafficking in Human Beings in EU Member States and Candidate Countries. As presented at the European Conference on Preventing and Combating Trafficking in Human Beings. September 18-20. Brussels, Belgium. Text available at http://www.belgium.iom.int/STOPConference/Confdocs/index.htm

Holsopple, Kelly (1998). 'Stripclubs According to Strippers: Exposing Workplace Sexual Violence', Metropolitan Coalition Against Prostitution, Minnesota USA. Text available at http://www.uri.edu/artsci/wms/hughes/stripc1.htm

Holt, Kate (2002). 'Once they were girls. Now they are Slaves'. The Observer. February 3.

Homohan and O' Connell(1997). "contrasting monetary regimes in Africa", IMF working paper WP/97/64 may.

Horold Kootz Cryril (1980). Selection and Training for Entrepreneurship Development" ILO, Geneva, 1983.

http://dx.doi.org/10.1016/0165-1889(88)90041-3

http://www.blackwell-synergy.com/doi/pdf/10.1111/1467-8268.00051 (accessed 05.11.2007

http://www.ilo.org/washington/areas/elimination-of-forced-labor/WCMS_182004/lang—en/index.htm.

http://www.infochain.org.htlm (Accessed on 26 September 2003).

http://www.nytimes.com/2014/02/01/opinion/the-super-bowl-of-sex-trafficking.html.

http://www.nytimes.com/roomfordebate/2012/04/19/is-legalized-prostitution-safer/criminalize-buying-not-selling-sex

http://www.pbs.org/unforgivableblackness/knockout/mannact_text.html.

http://www.state.gov/j/tip/rls/tiprpt/2007/.

Huang Y., Henderson G. E., Pan S. and Cohen M. S. (2004). HIV/AIDS risk among brothel-based female sex workers in China: assessing the terms, content, and knowledge of sex work. Sex Transm Dis, 31, 695-700.

Huber PJ (1964).'Robust estimation of a location parameter', Annals of Mathematical Statistics, 35(1):73–101.

Huber, P., (1964). "Robust Estimation of a Location Parameter," Annals of Mathematical.

HUDA S. (2006). Sex trafficking in South Asia. *Int J Gynaecol Obstet,* 94, 374-81. 23.

Hughes, Donna M (2001). Globalization, Information Technology, and Sexual Exploitation of Women and Children . Rain and Thunder - A Radical Feminist Journal of Discussion and Activism, Issue #13. Text available at http://www.uri.edu/artsci/wms/hughes/mhvlegal.htm

Hughes, Donna M. (1999). Legalizing Prostitution Will Not Stop the Harm, The Coalition Against Trafficking in Women, February. Text available at http://www.uri.edu/artsci/wms/hughes/pubvio.htm

Hughes, Donna M. (1999). The Fact book on Global Sexual Exploitation - Italy. Coalition Against Trafficking in Women. Text available at http://www.uri.edu/artsci/wms/hughes/italy.htm

Hughes, Donna M. (1999). The Harm of Sexual Exploitation - From the Individual to the State. Coalition Against Trafficking in Women. Text available at http://www.uri.edu/artsci/wms/hughes/demand.htm

Hughes, Donna M. (2000). Control and Abuse of Women and Girls' Sexuality, The Coalition Against Trafficking in Women Text available at http://www.uri.edu/artsci/wms/hughes/pubvio.htm

Hughes, Donna M. (2000). Global Sexual Exploitation—Supply and Demand Markets. University of Rhode Island, USA. Text available at http://www.uri.edu/artsci/wms/hughes/pubvio.htm

Hughes, Donna M. (2000). Men Create the Demand; Women Are the Supply, Lecture on Sexual Exploitation. Queen Sofia Center, Valencia, Spain. November. Text available at http://www.uri.edu/artsci/wms/hughes/demand.htm

Human Rights Watch (2002). Refugees, Asylum Seekers, Migrants and Internally Displaced Persons: The Year in Profile. Text available at www.hrw.org/wr2k2/print.cgi?refugees.html

Human Rights Watch, Rape for Profit, Trafficking of Nepali Girls and Women to India's Brothels. New York: Human Rights Watch, 1995.

Human Rights Watch, Women's Human Rights Division (2002). 2001 Year in Review. New York, NY USA. Text available at http://www.hrw.org/wr2k2/women.html

Human Rights Watch, Women's Rights Division (2001). Recommendation regarding the Proposal for a Council Framework Decision on Combating Trafficking in Human Beings. New York, NY USA. Text available at www.hrw.org/backgrounder/wrd/trafficking-framework.pdf

Humphrey, J. and L. Navas-Aleman. (2010). Value chains, donor interventions and poverty reduction: A review of donors practices, University of Sussex, ID S Research Report 63.

Hussain, S. and Malik, S. (2011). "Inflation and Economic Growth: Evidence from Pakistan" International Journal of Economics and Finance, 3(5).

Hyndman, R.J., Koehler, A.B., Ord, J.K., and Snyder, R.D. (2005) "Prediction ... using two new classes of state space models," Journal of Forecasting 24: 17–37.

Hyndman, Rob J., Koehler, Anne B., Ord, J. Keith, and Ralph D. Snyder, (2005). "Prediction

Ibrahim, J. (2002). Notes on globalization and the marginalisation of Africa. CODESRIA Bulletin, 3 and 4.

IFAD. (2000). Gender perspective: focus on rural poor: An overview of gender issues in IFAD assisted projects, Programme Management Department. Rome: IFAD.

IFAD. (2011a). Access to Markets: Making Value Chains Work for Poor Rural People. Accessed online at http://www.ifad.org/englis h/market/index.htm on April 20, 2012.

IFAD. (2011b). Rural Poverty Report 2011. Rome: International Fund for Agricultural Development.

IFAD. (2012). Youth in Agriculture, Special Session of the Farmers' Forum Global Meeting, Synthesis of Deliberations, 18 February, Italy.

IFC Report on Small and Medium Enterprises, No. 203.

Ikuomola, Y. (2012)."Volunteering: standing up to be counted in the task of nation-building". Venture Career Journal. pp 92-93.

ILO (1995). "General Conditions to Stimulate Job-Creation in Small and Medium-sized Enterprises" Report V(1), International Labour Conference 85th Session 1997, ILO, Geneva, 1995.

ILO (1996). World Employment. www.ilo.org/public/english bureau/inf/pkits.

ILO/IMF/OECD/Eurostat/UN/World Bank, ILO/IMF/OECD/Eurostat/UN/World Bank.

ILO/IMF/OECD/UNECE/Eurostat/World Bank. 2004. Consumer Price Index Manual:

IMF (2002). World Economic Outlook International Monetary fund, Washington DC.

Implications The Nigerian banker, April-June, 6-8.in Nnadi, K.U. and Ogwude I.C (ed) Transportation in Nigeria: Economics, Policy and in South Africa," pp. 55–75 in Carson and others (2002). op. cit.

Indicator," Banco de Portugal Working Paper, WP 3-04 (January).

Indicators', in K. Lahiri and G Moore (eds.) Leading economic indicators: new Inflation Estimation," National Bureau of Economic Research Working Paper No. 6183.

Inflation for the Euro Area," German Economic Review, 4(3):269–306.

Inflation in Pakistan, State Bank of Pakistan – Research Bulletin, 1(1-2): 35-44.

Inflation Indicator," Banco de Portugal Working Paper, WP 9-01 (September). Iation Targeting," pp. 157–171 in Carson and others (2002) op. cit.

Inflation." Federal Reserve Bank of Cleveland Economic Review29, 15-24. Inflation: a weighted-median measure], mimeo, Prices Division, Statistics Canada. inflation: evidence and policy implications', IMF Staff Papers, 42(2):344-74. Institution, (Washington, D.C.).

International Journal of Forecasting, 5:559–583.

International Journal of Forecasting, 7:31–37.

International Journal of Business and Social Science, 3(10) Wai, U. T. (1959). "The Relationship between Inflation and Economic Development: A Statistical Inductive Study", IMF Staff Papers, 7: 302-317.

International Labour Office (ILO). (2004). Improving prospects for young women and men in the world of work: A guide to youth employment. Policy Considerations and Recommendations for the Development of National Actions Plans on Youth Employment. Accessed online at http://www.ilo.org/public/libdoc/ilo /2004/104B09_280_engl.pdf on April 23, 2012.

International Labour Office (ILO). (2007). African Employment Trends. Accessed online at http://www.ilo.org/wcmsp5/groups/public/—ed_emp/—emp_elm/— trends/documents/publication/wcms_114278.pdf on April 23, 2012

International Monetary Fund (2006). "Inflation Targeting and the IMF" by Nicolatia Batini, Peter Bever Kalpana. Kochhar and Scott Rogar. www.imf.org.

International Statistical Review, (1974). Part I, Vol. 42, No. 2, pp. 147–74; Part II, Vol.

Intervals for Exponential Smoothing Using Two New Classes of State Space Models,"

Isijola, CO., (2000). Impact of financial sector, Reform on the supply and Demand For Agricultural Credit in Nigeria. First bank BiOAnn. Rev., 8:16.

Iyoda M. (2005). "The Determination of Investment in monetary production economy: A theory and it Empirical Application in the post Keynesian Tradition". Journal of Post Keynesian Economics, Fall.

J. K Onoh, (1990). "The State of the Nigeria Economy: An Over-view of Structural Adjustment Programme in Nigeria: Impact of SFEM on the Nigeria Economy," Nigerian Institute of International Affairs, Lagos.

J.O. Osakwe, "Government Expenditure, Money Supply and Prices: (1970-1980). CBN Economic and Financial Review.

Jacobson, Tor and Karlsson, Sune, (2004). "Finding Good Predictors for Inflation: A Bayesian

Jaffee, S., S. Henson and L. Diaz Rios. (2011) . Making the grade: Smallholders Framers, emerging standards and development assi stance programs in Africa, Word bank, A research program synthesis, Report No. 62324-AFR

Jayedau, Arjun, (2006). "Dittering preferences between anti-inflation and anti-unemployment policy among the rich and poor: Economics Letters 91. Pp. 67-71.

Jayeday, Argun (2008). "The class content of preferences toward Anti-inflation and Anti-unemployment policies". International Review of Applied Economics.

Jayne, T.S., N. Mason, R. Myers, J. Ferris, D. Mather, M. Beaver, N. Lenski, A. Chapoto and D. Boughton. (2010). Patterns and trends in food staples markets in Eastern and Southern Africa: Toward the identification of priority investments and strategies for developing markets and promoting smallholder productiv ity growth, Michigan State University 33

Johansen, S. (1988). Statistical Analysis of Cointegration Vectors. Journal of Economic.

Johansen, S. and Juselius, K.(1990). Maximum Likelihood Estimation and Inference on Co-integration with the Application to the Demand for Money. Oxford Bulletin.

Johanson, H. G. (1967). Is Inflation a Retarding Factor in Economic Growth?in Fiscal and Monetary Problems in Developing States, Proceedings of the Third Rehoroth Conference, ed. By David Krivine

Jones, R.C. (1998). "Introduction: The Renewed Role of Remittance in the New Order", Economic Geography 74C17, 1-7.

Jorgenson, D.W. (1963). Capital theory and investment behaviour" American Economic Review, 1963, 53(2):247-59.

José, Armida San, Slack, Graham L., and Subramanian S Sriram, (2002). "Statistical Principles.

José, Ramos Maria, (2004). "On the Use of the First Principal Component as a Core Inflation

Journal of Agile management systems. 2(1), 40-53.

Journal of Humanities and Social Science Vol. 1 No. 18 www.ijhssnet.com

Journal of Money, Credit, and Banking, 37(2):245–72.

Jovanovic B. (2004). Asymmetric cycles NBER working paper.

Kaplinsky, Raphael and Morris, Mike. (2000). A Handbook for Value Chain Research (Ottawa: International Development Research Center). Kaplinsky, Raphael and Morris, Mike. (2003). Governance Matters in Value Chains. Developing Alternatives 9(1):11-18.

Kaplinsky, Raphael. (2000). Globalisation and Unequalisation: What Can Be Learned from Value Chain Research? Journal of Development Studies, 37(2):117-146.

Karnaiah, A., (2001). A Monetary Policy, expected inflation real activity and stock returns in India: An empirical analysis. Asia-Mr. J. Econometric, pp: 191-200.

Katarzyna Karpowicz, (2008). "Economic Growth and Development: Is High GDP Enough? Aberystwyth University.

Katz and Kahn (1966). Keynote on small and medium enterprises; Issues and Prospects.

Kearns J (1998).'The distribution and measurement of inflation', Reserve Bank of Australia Research Discussion Paper No 9810.

Kearns, J., (1998). "The Distribution and Measurement of Inflation, Reserve Bank of Australia.

Keith and Guberlling C.E. (1975). Report on the Contemporary Issues on SME.

Kelecki, M. (1971). Selected Essays on Dynamics of Capitalist Economy 1933-1970, Cambridge: Cambridge University Press.

Keynes JM (1930). A Treatise on money, Harcourt Brace and Co, New York.

Keynes, J.M. (1936). The General Theory of Employment, Interest and Money Volume III of the Collected Writings of John Maynard Keynes, London: McMillan.

Khan, M. and C. Reinhart (1990). "Private Investment and Economic Growth in Developing Countries", World Development Vol. 18 (January):19-27.

Khan, M. S. and Senhadji, A. S. (2001). "Threshold Effects in the Relationship between Inflation and Growth", IMF Staff Papers, 48.

King, R.G. and R.E. Levine (1993). "Finance and Growth: Schumpeter might be right", Quarterly Journal of Economics Vol. 108 (August):717-737.

Kingsbury, D. (2010). Issues Paper on Value Chains for Opportunities and Growth: Roles for IFAD Co-Funded Projects. Regional Project Implementation Workshop in Western and Central Africa. November 8-11, Dakar, Senegal.

KIT, Faida MaLi, and IRR. (2006). Chain Empowerment: Supporting African Farmers to Develop Markets. Amsterdam: Royal Tropical Institute.

Kleinberg, S. and Campbell, R. (2008). Business Enabling Environment and the Value Chain,

Knorringa P., and L. Pegler (2006). Globalisation, Firm Upgrading, and Impacts on Labour. Royal Dutch Geographical Society KNAG, 97 (5): 470-479.

Koch, L. T. (2003). Theory and Practice of Entrepreneurship Education: A German view. In International Journal of Entrepreneurship Education, 1: 633-660.

Korsah. Kweku Baa, Ernest K. Nyarko and Noel A. Tagos, (2001). "Impact of Financial Sector Liberalisation on Competition and Efficiency in the Ghanaian Banking Industry". IFLIP Research paper 01-2 Geneva: ILO.

Kose, M.A., Pasad, E.S., Terrones, M.E. (2003). Financial integration and macro-economic isolatility. IMF staff paper 50:119-142.

Kose, M.A., Pasad, E.S., Terrones, M.E., (2005). How trade and financial integration affect relationship between growth and volatility? IMF working paper 05/19.

Krueger, N. F. (2000). The cognitive infrastructure of opportunity emergence. In Entrepreneurship: Theory & Practice, Spring Pg 5-23. 29.

Kurz, H. (1986). "Normal position and capacity utilization" political economy: Studies in the surplus Approach, (2):37-54.

Kwanashie, M. (1999). Concepts and dimensions of globalization, in Nigerian Economic Society, Globalization and Nigeria's economic development, Ibadan: NES.

Laflèche T (1997).'Statistical measures of the trend rate of inflation', Bank of Canada Review, Autumn, pp 29–47.

Laflèche, T (1997a). 'Mesures du taux d'inflation tendanciel' [Measures of trend inflation].

Laflèche, T (1997b). 'Statistical measures of the trend rate of inflation', Bank of Canada

Lafléche, T., (1997). "Statistical Measuresof the Trend Rate of Inflation," Bank of Canada

Laidler, D, and M Parkin (1975). 'Inflation: a survey', The Economic Journal, vol. 85.

Lambert, D.M. (2001). Supply Chain Management: What does it involve?

Lambert, D.M. and Stock, J.R. Ellram, L.M. (1998). Fundamental of Logistics Management:

Largley, K.M. (1968). "The External Resource Factor in Nigeria's Economic Development". Nigerian Journal 182 of Economic and Social Studies Vol. 10(2), pp. 151-190.

Lauret, F. (1993). Sur les études de filières agro-a limentaires. Economies et Society 17 (May).

Laxton, D, G Meredith and D Rose (1995). 'Asymmetric effects of economic activity on Layount = article webzine & artcleid = CA139528

Lehohla, Pali J. and Annette Myburgh, (2002). "Statistical Implications of Inflation Targeting.

Leiderman, L, and L Svensson, eds. (1995). Inflation targets, Centre for Economic Policy Lesotho.

Lessons from the International Experience. Princeton: Princeton University Press. Letters, 58:143–147.

Levuve, R. (2002). Bank- based or market-based financial system, which is better? Journal of Financial International (UCH):398-428.

Lewis, M. (1979). "Does Political Instability in Developing Countries Affect Foreign Investment Flow?: An Empirical Examination", Management International Review, 19:59-68.

Ley, D., (2006). Explaining variations in business performance among immigrant entrepreneurs in Canada. In Journal of Ethnic and Migration Studies, 32(5):743-764.

Li, M. (2005). "Inflation and Economic Growth: Threshold Effects and Transmission Mechanism" Department of Economics, University of Alberta, Canada (Mimeograph).

Linan, F. (2008). Skill and Value Perceptions: how do they affect entrepreneurial intentions? In International Entrepreneurship Management Journal, 4: 257-272.

Lipalile, M. (2001). Globalization, regional integration and development in Southern Africa, in Eddy Maloka (ed), A United States of Africa, Pretoria: Africa Institute of South Africa.

Loayza, Norman, and Raimundo Sato, eds., (1999). Inflation Targeting: Design, Performance.

Lucas, Robert E. (1976). "Econometric Policy Evaluation: A Critique," Carnegie-Rochester

Lundy, Mark, María Verónica Gottret, Carlos Oste rtag, Rupert Best, and Shaun Ferris. (2007). Participatory Market Chain Analysis for Sma llholder Producers. Cali, Colombia: CIAT.

Lunn, J. (1980). "Determinants of US Direct Investment in the EEC: Further Evidence", European Economic Review, Vol. 13, pp.93-101.

Lusby, Frank and Panliburton, Henry. (2002). Subsector/Business Service Approach to Program Design. Office of Microenterprise Development, USAID. Washington, DC: USAID.

Lyoha M.A. (1998). Macroeconomics for Developing World Benin City: Miyo Educational Publisher.

M. A. Adebiyi, "Puin Nigeria: An Autoregressive Model," (2006). www.isser.org/43%20 Adebayo.PDF

M. A. Onitiri and K. Awosika, (1982). "Inflation in Nigeria," Proceedings of a National Conference convened by the Nigerian Institute of Social and Economic, Ibadan.

M. Iyoha, (2002). "Budget of Economic Growth and Development," Bullion Publication of CBN, Vol. 26(2):12-27.

M. Z. Abdul Majid, (2007). "Causality Link between Money, Output and Prices in Malaysia: An Empirical Re-Examination," Applied Econometrics and International Development, Vol. 7, No. 1:81-93.

Machado JF, CR Marques, PD Neves and AG da Silva (2001).'Using the first principal component as a core inflation indicator', Banco de Portugal Working Paper No 9-01.

Machado, José Ferreira, Marques, Carlos Robalo, Neves, Pedro Duarte, and Macklem, Tiff, (2001). "A New Measure of Core Inflation," Bank of Canada Review, Vol. 3.

Maddala, G. S. (1992). Introduction to Econometrics, New York: Macmillan Publishing Co.

Madunagu, E. (1999). Globalization and its victims, in The Guardian, Lagos.

Magistro, John, Roberts, Mike, Yoder, Robert , Kramer, Fritz and Weight, Elizabeth (2004). PRISM: A Model for Smallholder Wealth Creation through Integrated Service Provision, Value Chain Market Development and Micro Irrigation Technology. Paper presented at the IWMI Regional Workshop and Policy Roundtable on Pro-Poor Intervention Strategies in Irrigated Agriculture in Asia, 25-27 August, Colombo, Sri Lanka.

Malinvaud, E. (1980). Profitability and Unemployment. London: Cambridge University Press.

Malla, S. (1997). "Inflation and Economic Growth: Evidence from a Growth.

Mallik, G. and Chowdhury, A. (2001). Inflation and Economic Growth: Evidence from South.

Malnvaud, E. (1982). "Wages and Unemployment", Economic Journal, 92(1) Management. Reliable Publisher, Owerri, Nigeria.

Mankikar A and J Paisley (2002).'What do measures of core inflation really tell.

Mankikar, Alan and Jo Paisley, (2004). "Core Inflation: A Critical Guide," Bank of England.

Mankiw, N. Gregory and Ricardo Reis, (2003). "What Measures of Inflation Should a Central manufacturing companies in Southwestern Nigeria, (2002-2006), Unpublished Ph.D Thesis.

Maria JR (2004).'On the use of the first principal component as a core inflation indicator', Banco de Portugal Working Paper No 3-04.

Market Brief, April 20, (2011). Tunis: African Development Bank.

Marques CR, PD Neves and AG da Silva (2002).'Why should central banks avoid the use of the underlying inflation indicator?', Economics Letters, 75(1): 17–23.

Marques CR, PD Neves and LM Sarmento (2000).'Evaluating core inflation.

Marques, Carlos Robalo, Neves, Pedro Duarte, and Sarmento, Luís. Morais, 2000.

Marques, Carlos Robalo, Neves, Pedro Duarte, and Sarmento, Luís. Morais, 2003.

Matthias, L. and Herr Tapera, J. (2009). Value Chain Development for Decent Work: A guide for development practitioners, government and private sector initiatives, ILO.

Mckinley Temy (2006). "the monopoly of Global capital flow: who needs structural Adjustment Now?, UNDP, international poverty center, working paper ₦12.

McKinnon, R.I. (1973). Money and Capital in Economic Development, Washington DC: The Brookings Institutions."

Measure of Inflation," Federal Reserve Bank of Cleveland Economic Review, (4):15–24.

Measurement and Modeling," in The Future of Inflation Targeting, ed. by Christopher.

Measurement considerations', Bank of Canada Review, September, pp. 3-23.

Measurement of Price Indexes." In Scanner Data and Price Indexes, ed. M.D. Shapiro.

Measures," De Nederlandsche Bank NV, Research Memorandum WO&E No. 639/0034.

Measuring Cost-of-Living and Price Indexes. Washington: National Academy Press.

Meredith, O. (1983). The Practice of Entrepreneurship. Geneva : International Labour Office.

Meyer-Stamer, Jorg and Waitring, Frank. (2006) . Value Chain Analysis and Making Markets Work for the Poor: Poverty Reduction Through Value Chain Promotion.

Meyer-Stamer, Jorg. (2002). PACA: Participatory Appraisal of Competitive Advantage http://www.paca-online.de.

Micro Links Wiki. (2010a). Youth and the Value Chain Approach. Accessed online at http://apps.develebridge.net/amap/inde x.php/Youth_and_the_Value_Chain_Approach on April 18, 2012.

MicroLinks Wiki. (2010b). Value Chai n Approach. Accessed online at http://apps.develebridge.net/amap/inde x.php/Value_Chain_Approach on April 18, 2012.

Miehlbrade, Alexandra O., McVay, Mary and Tanburn, Jim. (2006). Implementing Sustainable Private Sector Development : Striving for Tangible Results for the Poor. Geneva : International Labour Office.

Miehlbradt, Alexandra O., McVay, Mary and Tanburn, Jim. (2005). From BDS to Making Markets Work for the Poor: the 2005 Reader (Gen eva: International Labour Office).

Miligan, B. (2000). Transportation and the Supply Chain: Transportation still falls through mimeo, Oesterreichische National bank.

Minsky, H. Joun Mayn\ard Keynes, New York: Columbia, 1975.

Minten, B., L. Randrianarison and J. Swinnen. (2005). Global re tail chains and poor farmers: Evidence from Madagascar, LICOS Discussion paper 16406, Centre for Institutions and Economic Performance, UK.

Mishkin, F. (1992). The Economic of money, Banking and Financial Markets. 3rd Edn. Harper Collin Publishers New York.

Mishkin, Frederic, S. and Schmidt Hebbel (2007). Does inflation Targeting make a difference? HBER paper, 12876.

Mishkin, Trederic S. and Klaus S. Schmidst-Hebbel (2001). One Decade of inflation: Targeting in the world: what do we know and what do we need to know? NBER working paper 8397.

Mitchell, J., J. Keane and C. Coles. (2009). Trading up: How a Value Chain Approach can benefit the rural poor, COPLA - Comercio y Pobreza en Latino America and the Overseas Development Institute, UK. 34

Mittleman J.H. (1996). How does globalization work? in Mittleman J. H (ed), Globalization: Critical reflection, Boulder, Colorado.

Mittleman, J. (2000). The Globalization syndrome: Transformation and resistance, Princeton: Princeton University Press.

Model Averaging Approach," Journal of Forecasting, 23: 479–496.

Mohanty, D, Chakraborty, A. B., Das, A. and John, J. (2011). "Inflation Threshold in India: An Empirical Investigation" Reserve Bank of India Working Paper Series, 18.

Monetary Economics, 14(2):251–64.

Monetary Policy, ed. by Gregory N. Mankiw (Chicago: University of Chicago Press for NBER), pp. 195–215.

Money, Credit, and Banking 5:173-191.

Montigard, J. (1992). L'analyse des filières agro-A limentaires. Economies et Sociétés, Cahiers de l'ISMEA 26(6), pp. 59-70.

Montmarquette, Ottawa: Statistics Canada, 1049-1113.

Mukhejee, C, (2002). The nature of the Causal relationship Between Stock Market acroeconomic Aggregates in India: An Empirical Analysis. Panda, Chakradhara Department of Economics, Jadavpur University.

Nabudere, D. (2000). Globalization, the African post-colonial state, post traditionalism and the new world order, in D. Nabudere (ed), Globalization and the post-colonial African state, Harare: AAPS Books.

Nabudere, Dani W. (1997). 'Beyond modernization and development, or why the poor reject development', Geografiska Annaler, Series B: Human Geography, http://www.blackwell-synergy.com/doi/pdf/10.1111/j.0435-3684.1997.00019.x (accessed 10 November 2007).

National Bureau of Economic Research Working Paper4168.

National Bureau of Statistic (NBC) Nigeria Poverty Profile Report, (2010). Press Briefing by the Statistician-General of the Federation, 13th Feb. 2012.

National Bureau of Statistics (2005). The Nigerian statistic al fact sheets on Economic and Social Development, FOS, Nigeria.

National Development Plan (1975- 1980). Prepared by Nigeria Government Expertise.

National Planning Commission (2004). National Economy and Empowerment Strategies (NEEDS) Abuja, Nigeria.

NDIC, (1996). Nigeria Deposit Insurance Corporation Annual Reports and Accounts.

Ndikumana Leonce (2000). "Financial Determinations of Domestic Investment in Sub-Saharan Africa. Evidence from panel Data". World Development, 28(2): 381- 400.

New Zealand Discussion Paper No. G98/9 (July), Wellington, New Zealand.

New Zealand. Discussion Paper G97/7 Wellington, New Zealand.

Ngigi, Margaret et al. (2010). Small holder dairying in Eastern Africa. Chapter 6 in Steven Haggblade and Peter B.R. Hazell (editors) Successes in African Agriculture: Lessons for the Future. Baltimore: Johns Hopkins University Press.

Ngozi Okonjo-Iweala and Philip Osafo-Kwaako (March 2007). Nigeria's Economic Reforms: Progress and Challenges The Brookings Institution 1775 Massachusetts Ave., NW Washington, DC 20036.

Ninsin K. A (2000). Globalization and the future of Africa. *AAPS Occasional paperseries*, 4, (No.1).

Nissanke maehiko and Aryeeted, Ernest (1998). financial integration and development Liberalization and Reform in sub-saharan Africa, London and New-York : Routledge.

Nissanke, Machiko and Aryeetey, Ernest (1998). Financial Integration and Development: Liberalization and Reform in Sub-Saharan Africa. London and New York: Routledge.

NIU Outreach (2005). The Role of Higher Education in Economic Development http://www.hearrr.org/pdf/HEARRRWhitePaper.pdf

Nnanna, O.J and Dago, Mela (1998). structure Reform, monetary policy and financial deeping: The Nigerian experience, economic and financial Review, vol.36 no.2, June pp. 1-29.

Nwangwu, I.O. (2007). Higher education for Self Reliance: An imperative for the Nigerian economy. NEAP publication pg. 1-8.

Nzott, S.M (2004). money, banking and finance, theory and practice Owerri. Huason Jude Publisher.

Obadan, M.I. (1994). "Direct Investment in Nigeria: An empirical Analysis", African Studies Review 25(1):67-81.

Objectives and operations?', The Economic Journal, 104:1424-36.

Oboh, G.A.T., (2005). <a Globalization and Competitiveness; concept, Fears and Prospect For Common Wealth Countries. Union Digest, 9:384.

Ocampo, Jose Antonio (2002). "Capital Account a Counter-Cyclical Prudential Regulations in Developing Countries". WIDER Discussion Paper, Helsinki WIDER UNU.

Ocampo, Jose Antonio (2003). A capital Account and counter-cyclical prudential Regulation in Developing countries" in form capital surges to Drought. Seeking stability for emerging Economic, edited by Richardo French –Dauis and stephany Gritith – Jones 217-44, New –York, Palgrava Macmillan. Occasional Paper, No. 202, Washington D.C., IMF.

Odozi, V.A. (1995). "An Overview of Foreign Investment, 1960-65" CBN Research Department Occasional Paper, No. 11 (June).

OECD (Organisation for Economic Co-operation and Development) (2005). 'Measuring and assessing underlying inflation', OECD Economic Outlook, 77(1):322–343.

OECD Statistics Working Papers No. 2005/5, Organisation for Economic Co-operation.

OECD. (2006). Promoting Pro- Poor Growth. Policy Statemen t. DAC High-Level Meeting of Australia Bulletin, October, pp 1–5.

Ofei, Kwadwo Ansah (2001). "Terms and Access to credit: Perceptions of SME/ Entrepreneurs in Ghana", University of Nigeria.

Office of the Secretary to the Government of the Federation (OSGF) (2012). Nigeria in Transformation: Executive Intelligent Management Course 9.

Ogamba, E.N. (2003). "The Impact of Globalisation on Foreign Capital Flows in Nigeria", An Unpublished Ph.D. Thesis submitted to the Department of Banking and Finance, UNN, Enugu Campus, Enugu State.

Ogun, T. P. and P.O. lyoha, (2005). The Nigerian Stock Market and Future Economic Activity: Does the Deregulation of Financial Markets make any Difference? Union Digest, 1:2.

Ogunleye, G. A., The causes of Bank Failure and Persistent Distress in the Banking Industry. NDIC Quarterly, Vol. 13.

Okagbue. S. and T, Aliko, (2005). Banking Sector Reforms in Nigeria International legal News, 2:2.

Okun, A (1970). 'Inflation: the problems and prospects before us', in A Okun, H Fowler Olabisi Onabanjo University, Ago-Iwoye.

Olakampo, O. (1962). "Foreign Aid and the Plan", Nigerian Journal of Economic and Social Studies, 492:116-125.

Olaloku, F.A. (1979). Structure of the Nigerian economy, London.

Olaniyi, O. (1988). "An Econometric Analysis of Domestic Savings and Investment Financing in Nigeria", Journal of Teacher Education, 4(1):133-142.

Olofin, S.O. and Afandiegin, U.J. (2008). Financial structure and economic growth in Nigeria Journal of Security and finance, 13(1):47-48.

Olopoenia, R.A. (1983). "Foreign Investment and Growth Rates of Capital -Importing Countries: A Technical Note", Nigerian Journal of Economic and Social Studies, Vol. 25, No. 2 (July) pp. 221-232.

Olowoyeye(1986). A keynote on small and medium Enterprises in Nigeria.

Omolayo B. (2006). Entrepreneurship in Theory and Practice. I n F. Omotosho, T.K.O. Aluko, O.I. Wale Awe and G. Adaramola (eds). Introduction to Entrepreneurship development in Nigeria. Ado-Ekiti; UNAD Press. on the CPI', Reserve Bank of New Zealand Research Note 96/1.

Onyekpe, N.J.G. (2001).Globalization and liberalization of the world economy: Concepts, historical development and implications for the less developed countries and labour movement, in Sylvester Odion-Akhaine (ed), Globalization, United Nationsand the fringe players, Lagos: CENCOD. Operation, New Jersey; Upper Saddle River.

Oriakhi, F. (2001). Globalization as imperialism : The Nigerian experience, in Sylvester Odion-Akhaine (ed), Globalization, United Nations and the fringe players, Lagos: CENCOD.

Osaghae, E. (1998). Crippled giant: Nigeria since independence. London: C. Hurst and Company Publishing Limited.

Oshikoya, T. (1994). "Macroeconomic Determinants of Domestic Private Investment in Africa: An Empirical Analysis" Economic Development and Cultural Change, Vol. 42(3).

Oxford Bulletin of Economics and Statistics, 48: 213-228.

Oyaide, W.J. (1977). The Role of Direct Private Foreign Investment in Economic Development: A Case Study of Nigeria, 1962-1973. Washington DC: United Press of America.

Oyesiku, O.O. (1995). Understanding Multiple Regression Analysis. Lagos: HEBS Publisher.

Pakes, A. (2002). "A Reconsideration of Hedonic Price Indices with an Application to Paper.

Parks, R (1978). 'Inflation and relative price variability', Journal of Political Economy.

Paul, S., C. Kearney and Chowdhury K. (1997). Inflation and economic growth: a multi-country empirical analysis. Applied Economics, 29:387-1301.

PCs." Working Paper No. 8715, National Bureau of Economic Research, Cambridge.

Philips and Titilola, (1997). Vision 2010, NISER. Physical Distribution and Material Management, 7(3): 19-33.

Pittman, K. (1993)."About promoting youth development in schools". Seminar with 4-H Faculty, St. Paul, University of Minnesota.

Policy, ed. N.G. Mankiw, Chicago: University of Chicago Press, 195-215. Political Economy, 96(6):1303–1314.

Pollak, R.A. (1989). "The Treatment of the Environment in the Cost-of-Living Index." Pollin, R. and Zhu. Inflation and economics Growth: Across country Non-Linear Analysis. Journal of post Kiymesian Economics summer 40128 No.4.

Pollin, Robert and Andong Ahu. (2005). "Inflation and economic growth: a cross-country non linear analysis, "PERI Working paper 109, Amherst, MA: political Economy Research Institute.

Porrado, Eric and Andrés Velasco, (1999). "Alternative Monetary Rules in an Open Economy.

Porter, Michael E. (1998). Clusters and the New Economics of Competition. Harvard Business Review 76(6):77-90.

Portugal, Economic Bulletin, (1997). Practical Issues for Emerging Market Countries," International Monetary Fund (IMF), Prices Division, Statistics Canada, 3(1):1130-44.

Proctor F. and Lucchesi V. (2012). Small-Scale Farming and Youth in an Era of Rapid Rural Change. Knowledge Programme Small Produ cer Agency in the Globalised Market, International Institute for Environment and Development (IIED), London.

Quah D and SP Vahey (1995).'Measuring core inflation', Economic Journal, 105(432): 1130–1144.

Quisumbing, A.R. and Maluccio, J.A. (2000). Intrahousehold allocation and gender relations: New empirical evidence from four deve loping countries, International Food Policy Research Institute, Food Consumption and Nutrition Division, Discussion paper No. 84, USA.

R. A. Ajisafe, (1996). "An Empirical Investigation of the Determinants of Inflation in Nigeria," Journal of Economics and Finance, 3:65-94.

Rabiu, M. (2012)."Promoting ethics and organizational values in the work place". Venture Career Journal. pp 94-95.

Ramey, G. and Ramey V.A. (1995). Cross-country evidence of the link between volatility and growth. American Economic Review 85:1138-1151 (December.).

Rationing Model with Monopolistic Competition", European Economic Review, 3(2) National Bureau of Economic Research.

Raymond, J.L., Maroto, J.A. and Melle, M. (1999). Inversion Crecimiento empresial: Factors conducionantes. "Peoples de Economia Espanola, 78-79:102-121.

Real Output Change." In Price Level Measurement, ed. W.E. Diewert and C. Reinhart Carmen and kennetti Rogott (2008a). "This time is different. A panoramic view of Eight centuries of financial crises "NBER working papei 13882, March.

Reinhart, Carmen and Kcenneth Rogott (2008b). "Banking Crises: An equal opportunity menace". NBER, Working paper on: 14587. December.

Reserve Bank of Australia (1998). 'The implications of recent changes to the consumer price index for monetary policy and the inflation target', Reserve Bank.

Reserve Bank of New Zealand (1996). Briefing on the Reserve Bank of New Zealand, Review (Autumn), pp. 29-47.

Robinson, J.V. (1962). Essay on the theory of Economic Growth, London, Macmillan. Rochester Conference on Public Policy on the State of Economics.

Roger S (1997).'A robust measure of core inflation in New Zealand, 1949–96', Reserve Bank of New Zealand Discussion Paper No G97/7.

Roger S (1998).'Core inflation: concepts, uses and measurement', Reserve Bank of New Zealand Discussion Paper No G98/9.

Roger, S (1995). 'Measures of underlying inflation in New Zealand, 1981-95', Reserve

Roger, S (1996). 'The 1986-87 and 1989-90 GST shocks: revised estimates of the impact.

Roger, S (1997). 'A Robust measure of core inflation in New Zealand, 1949-96', Reserve

Roger, Scott, (1997). "A Robust Measure of Core Inflation in New Zealand," Reserve Bank of New Zealand.

Roger, Scott, (1998). "Core Inflation: Concepts, Uses and Measurement," Reserve Bank of New Zealand.

Rogers, Simon; Sedghi, Ami (15 April 2011). "How Fitch, Moody's and S&P rate each country's credit rating". *The Guardian*. http://www.guardian.co.uk/news/datablog/2010/apr/30/credit-ratings-country-fitch-moodys-standard. Retrieved 31 May 2011.

Rowlatt, Amanda, (2001). "The U.K. Office for National Statistics and the Inflation Target".

Rudebusch, Glenn D., (2005). "Assessing the Lucas Critique in Monetary Policy Models".

S. A Laryea and U. R. Sumaila, (2001). "Determinants of Inflation in Tanzania," Chr. Michelson Institute, CMI Work-ingPapers WP/2001/12 Bergen, Norway.

S. Olorunfemi, "Infrastructural Services and Manufacturing Growth in Nigeria: A Dynamic Analysis," Journal of Economics, Vol. 6, No. 2,David E. Lebow and Jeremy B. Rudd 2006-43.

Salami, D. and I. Kelikume (2010). "An Estimation of Inflation Threshold for Nigeria (1970-2008)". International review of Business Research papers, 6(5): 375–385.

Salian, P. and Gopakumar, K.(2010). Inflation and Economic Growth in India: An Empirical.

Samuelson, P (1939). "Interaction between the multiplier analysis and principle of acceleration. "Review of Economic Statistics, 21 (2):75-78.

Sani, T.A. and R.F. Yakpogoro, (1997). Thc Impact of Banking reform on Commercial Banks Performance in Nigeria. J. Indus. Business.. (JTBER) 1:2.

Sannes, R., (2001). Self-Service Banking. Value Creation Models and Information. Exchange. Special Services on Information Exchange on Electronic Markets, 4:3.

Sarel, M. (1996). "Nonlinear Effects of Inflation on Economic Growth", IMF Staff Paper, 43(1): 199-215.

Scapelanda, A., and R. Balough (1988). "Determinants of US Direct Investment in the EEC", *European Economic Review,* 21:381-390.

Scarf, H. and Shoven, J.B. (1984). Applied General Equilibrium analysis. Cambridge University Press.

Schaechter, Andrea, Stone, Mark, R. and MarkZelmer, (2000). "Adopting Inflation Targeting:

Schmoth (1993). Small and Medium Enterprises Report.

Scholte, J. (2000). Globalization : A critical introduction, London: Macmillan Press.

Seabrook, Jeremy (1993). Victims of Development: Resistance and Alternatives. London, New York: Verso.

Seleteng, M (2004). Inflation model for Lesotho (Unpublished paper). Central Bank of Series with a Unit Root. Econometrica, 49:1057-1072.

Serrano, F. (1995). "Long period Effective Demand and the Scattian supermultiplier" contribution to political economy, 14:67-71.

Serven, L. and A. Salimano (1991). "Private Investment and Macroeconomic Adjustment: A Survey", World Bank Research Observer Vol. 7 No. 1.

Shaw, E. (1973). Financial Deepening in Economic Development, New York: Oxford University Press.

Shaw, E.S., (1973). Financial Deepening and Economic Development New York Oxford University Press.

Sheikh, A. (1991). "Wandering around the warranted path: Dynamic Non-linear solution to the harrodian Knife-edge" in E.J. Heil and W. Semmler (eds) Nicholas Kalder and mainstream Economics: London: Macmillian, pp. 320-334.

Shiratsuka S (1997).'Inflation measures for monetary policy: measuring the underlying inflation trend and its implication for monetary policy implementation'.

Shonekan (1992). A Keynote delivered at Faculty of Bussiness Administration Conference. Shortrun? Journal of Applied Economics, 4(1), 89-105.

Shonekan (1992). Review on Small and Medium Enterprises, Vol. 84(3): 433–444.

Shu C and A Tsang (2004).'Alternative measures of core inflation on the mainland', Research Memorandum, Hong Kong Monetary Authority, May.

Sidrauski, M. (1967). Rational Choice and Patterns of Growth in a Monetary Economy.

Silver, Mick and Christos Ioannidis, (1996). "Inflation, Relative Prices and their Skewness,"

Singu, T. (2009). "Testing the neoclassical long-run and the Keynesian short-run Effect of Investment on output and Growth in India. "Journal of post Keynesian Economics Winter 2008-2009, 31(2):271-297.

Skjott-Larsen, T. (2000). Third-Party Logistics: From an Interorganisational Point of View:

Smyth, D. J. (1992). "Inflation and the Growth Rate in the United States' Natural Output", Journal of Applied Economics, 24.

Smyth, D. J. (1994). "Inflation and Growth", Journal of Macroeconomics, 16.

Smyth, D. J. (1995). "Inflation and Total Factor Productivity in Germany", Weltwirtschaftliches Archiv 131: 403-405.

Sneessens, H. (1987). 'Investment and the Inflation - Unemployment Trade - Off in a Macroeconomic Social Measurement 4, 179-195.

Soludo, C.C. (2004). 13-Point Nigerian Bank Reform, Social Science, 3(10).

Somuyiwa, A.O (2007). "The Relationship between the concepts of Logistics and Distribution".

Sorensen, Alan T., (2000). "Equilibrium Price Dispersion in Retail Markets for Prescription.
Sowa, Nii K. (2002). "Financial Sector Reform Policies and Poverty Reduction, "Working paper Center for policy Analysis (CEPA), Accra, Ghana.
Spolander, M (1994). 'The indicator of underlying inflation: basic idea and use', Bank of Stability," pp. 137–156 in Carson and others (2002) op. cit.
Stallings, Barbara and Rogerio student (2006). "Finance for development". Washington, DC : Brooking institution and United Nation ECLAC.
Stank, T.P. and Goldsbug, T.J. (2000). A Framework for Transportation Decision Making in Statistical Association, 62:1179–86.
Stightz, Joseph, E. Manson, U.J. (1966). Financial markets public policy and East Asian miracles. World Bank Research Observer, U. (2):249-276.
Stightz,, Joseph, E., Manton, et (1966). Financial market, public policy and East Asian miracles, World Bank Research Observers 11(2):249-276.
Stigler, G.J., (1961). "The Economics of Information," Journal of Political Economy, Vol. 69.
Stock, J and M Watson (1991). 'A Probability model of the Coincident Economic.
Storey, Andy (2003). "Measuring Development", In McCann, Gerard and Stephen McCloskey (eds.) (2003). From the Local to the Global. Key Issues in Development Studies. London, Starling: Pluto Press. Studies, 60(4):889-902.
Symonides, J. (ed). (1998). Human Rights: New dimensions and challenges, Dartmouth, Dartmouth Publishing Company.
Szulc, Bohdan J. (1983). "Linking Price Index Numbers." In Price Level Measurement.
T. A. Oyejide, "Deficit Financing, Inflation and Capital Formation. The Analysis of the Nigerian Economy, (1957-1970)." Nigerian Journal of Economics and Social Studies, [1Vol. 14, No. 1, 1972, pp. 22-43.
Tandon, Y. (2000). Globalization and Africans options, in Dani W. Nabudere, (ed), Globalization and the post-colonial African state, Harare: AAPS.
Targeting in Practice: Strategic and Operational Issues and Applications to Emerging Market Economies(Washington: International Monetary Fund). Targeting ed. by Christopher Kent, and Simon Guttmann, pp. 140-60 (August), Reserve Bank.
The Quarterly Journal of Economics, February, pp. 161–193. The Supply Chain Crakcs: Available from: http://www.manufacturing.net/pur/index.asp?
The Theory of the Cost of Living Index, R.A. Pollak, New York: Oxford University Press.
Theory and Practice. Geneva: International Labour Office. Times and Pitman Publishing.
Titilola, S.O and Akanji B.O (2005). Strategies for Planning and Budgeting for Economic and Social Sectors in Nigeria. A Paper Presented At a Training Workshop on Seeds Planning and Budgeting Technique for Enhancing the Performance of Planning and Budgeting Officers in Lagos State.
Titilola, S.O. Managing the Nigerian Economy: Challenges of the National Economy and Empowerment Strategies (NEEDS). A Paper Presented At a Training Programme for the Nigerian Petroleum Development Company (NPDC) On Developing Managerial Skillto Asian Exchange Rates," Journal of Forecasting, (21):355–380.
Todaro, Michael P., Stephen C. Smith (2006). Economic Development. Harlow: Pearson Education Limited.
Tornell, A, WEster mann, F., Martinez, L. (2004). The positive work between financial liberalization, growth and crises. NBER Working Paper. 10293.
Toyo, E. (2000). Background to globalization. Ibadan: ASUU Publication series. UNDP, National human development reports for Nigeria, 2000/2001.
Trezzini, A .(1995). "Capacity utilization in long-run and Autonomous components of Aggregate Demand." Contributions to political Economy, 14:33-36.

Trezzini, A. (1998). "Capacity utilization in long-run: Some further considerations contributions to political Economy, 17:53-69.

U.S. Bureau of Labour Statistics. (2005). "The Consumer Price Index." In BLS Handbook.

Umaru, A. and A.A. Zubairu (2012). "Effect of Inflation on the Growth and Development of the Nigerian Economy: An Empirical Analysis".

UNDP (2007). MDG in Nigeria: Current Progress, UNDP, Abuja University Press). US?', Bank of England Quarterly Bulletin, 42(4):373–383.

USAID Washington, DC: Development Alternatives Inc. Bennell P. (2010). Investing in the Future: Creating Opportunities for Young Rural People. IFAD Paper, December.

Usman, S. (1999). Implications of globalization for the Nigerian economy, in Nigerian Economic Society, Globalization and Nigeria's economic development. Ibadan: NES.

Vavrus, J. and Fletcher, A. (2006). Guide to social change led by and with young people." The Freechild Project.http://www.freechild.org/socialchangeguide.htm

Vega, Juan-Luis and Mark A. Wynne, (2003). "A First Assessment of Some Measures of Core Vol. 105 (September), pp.1130–44.

Von Hagen, J (1995). 'Inflation and monetary targeting in Germany'.

Washington, DC: IFPRI. Dowds, Curtis M. and Hinojosa, Jorge. (1999). .An Introductory Guide to the Subsector Approach to Community and Economic Development Washington, DC: National Council of La Raza. Winter, 44(4):466.

Working Paper No. 11548, National Bureau of Economic Research, Cambridge, Massa.

Working Paper No. 242, pp. 1–36; Summary in Bank of England, Quarterly Bulletin, Working Paper 6183.

World Bank (1995). A Continent in Transition: Sub-Saharan Africa in the Mid 1990s. Africa Region, World Bank, 1995. Washington, D.C.

World Bank Institute (2008). Development Outreach, Vol. 10:3. www.ilo.org/public/english/bureau/stat/guides/cpi/index.htm.

Wynne MA (1997). 'Commentary for "Measuring short-run inflation for central bankers"', Federal Reserve Bank of St. Louis Review, 79(2):163–168.

Wynne MA (1999). 'Core inflation: a review of some conceptual issues', Federal Reserve Bank of Dallas Working Paper No 9903.

Wynne, M (1997). 'Commentary', Federal Reserve Bank of St. Louis Review (May/June).

Wynne, Mark A., (1999). "Core Inflation: A Review of Some Conceptual Issues.

Yaqub, N. (2003). Review of the 2000/2001 National human development report, in The Nigerian Social scientist, 6:1.

Yates, A (1995). 'On the design of inflation targets' in Haldane, pp. 135-69.

Yule, U., (1911). An Introduction to the Theory of Statistics, (11th Edition 1937).

Yusuf, A. A. (1994). Africa : A Case for Mass Democratic Participation, in Omoruyi, Omo *et al*, Democratization in Africa : African Perspective. Abuja: C.D.S.

NIPA Publications on Economics and Extension

S.No.	Title	Author	ISBN	Year
1	Agricultural Marketing: Perspectives and Potentials	Bhat, Anil	9789385516153	2016
2	Advances and Challenges in Agricultural Extension and Rural Development	Rathakrishnan,T.	9789380235035	2009
3	Agricultural Extension: Worldwide Innovations	Saravanan, R.	9788189422967	2008
4	Agricultural Marketing: Perspectives and Potentials	Bhat, Anil	9789385516153	2016
5	Agrobiodiversity and Sustainable Rural Development	S.K. Soam	9789383305896	2015
6	Agro-enterprises for Rural Development and Livelihood Security	Sharma J.P. & S.K.Dubey	9789380235851	2011
7	Agroforestry Systems for Resource Conservation and Livelihood Security in Lower Himalays	Panwar,P & Dadhwal	9789381450215	2012
8	Agroforestry: Principles and Practices	Patra, Alok Kumar	9789381450765	2013
9	Agroforestry: Systems and Practices	Puri, Sunil & Pankaj Panwar	9788189422622	2007
10	Agroforestry: Systems and Prospects	C.B.Pandey & O.P.Chaturvedi	9789381450970	2014
11	Ancestral Knowledge in Agri-Allied Science	Saha, Ratan Kumar	9789383305216	2014
12	Basics of Research Methodology	Imam,Ekwal	9789383305438	2015
13	Biotechnology in India: Initiatives and Accomplishments	Niladri Bag	9789385516252	2016
14	Biotechnology in India: Initiatives and Accomplishments	Niladri Bag	9789385516252	2016
15	Computers in Agriculture: Fundamentals and Applications	Sharma Manish & Anil Bhatt	9789385516160	2015
16	Conducting An Effective and Successful Training Programme	Sontakki,Bharat S.	9789383305223	2015
17	Dimensions of Extension Education	Mohapatra, B.P.	9789381450987	2014
18	Disasters: Strengthening Community Mitigation and Preparedness	Khanna, B.K. & Neena Khanna	9789380235455	2011
19	Economics,Marketing and Sales of Agricultural Products	Thakur S. Nath	9789381450659	2013
20	Efficiency Indices for Agriculture Management Research	Devasenapathy, P.&B.Gangwar	9788189422882	2008
21	Evaluation and Impact Assessment of Technologies and Developmental Activities in Agriculture,Fisheries and Allied Fields	Roy, A.K.	9789380235400	2011
22	Extension Education Management in Veterinary Sciences and Animal Husbandry	Singh, Ruchi	9789385516054	2016
23	Extension Management in the Information Age Initiatives and Impacts	Philip, H. & T.Rathakrishnan	9789381450543	2013
24	Extension of Technologies: From Labs to Farms	Anandaraja, N.	9788189422837	2008
25	Extension Techniques for Livestock Development	Singh S.K. & S.D.Singh	9789380235233	2011

S.No.	Title	Author	ISBN	Year
26	Family Farming and Rural Economic Development	M.L.Choudhary & Aditya	9789383305858	2015
27	Farm Women Empowerment Through Dairy Co-operative Societies	Swain, Pitambar	9789383305186	2014
28	Hill Agriculture: Economics and Sustainability	Sharma, Pawan & Sudhakar Dwivedi	9789381450871	2014
29	Human Resource Development	Sannigrahi, A. Kr.	9789380235769	2011
30	ICTs for Agricultural Extension: Global Experiments, Innovations and Experiences	Saravanan, R. ed.	9789380235240	2010
31	ICTs for Transfer of Technologies: Tools & Techniques	Verma, S.R.	9788193014479	2015
32	Information and Communication Technology for Agriculture and Rural Development	Saravanan, R.	9789380235882	2011
33	Information and Knowledge Management: Tools, Techniques and Practices	Roy, A.K.	9789381450628	2013
34	Information Technology in Veterinary Science	Patil, Vivek	9788190851244	2009
35	Innovations in Agribusiness Management	Lokanadan, Kornam	9789380235097	2009
36	Management of Agricultural Inputs	Mahapatra, S.	9789381450642	2013
37	Mobile Phones for Agricultural Extension	Saravanan,R.	9789383305230	2014
38	Question Bank in Extension Education	Mohammad,Asif	9788193014493	2015
39	R Statistics	Samuel, Duleep	9789385516146	2015
40	Rural Livelihood and Food Security	Wani, M.H.	9789380235936	2012
41	Samekit Krishi Pranali	Kumar, Sanjeev	9789381450154	2012
42	SHGs in Techno-Economic Empowerment of Tribal Women	Satapathy, C. & Sabita Mishra	9789380235875	2011
43	Social-Ecological Diversity and Traditional Food Systems	Singh Ranjay, Nancy J.Turner	9789383305360	2014
44	Technologies for Livelihood Enhancement	V. L. Chopra	9789383305810	2015
45	Traditional Agricultural Practices: Applications and Technical Implementations	Rathakrishnan,T.	9789380235028	2009
46	Women in Agriculture and Rural Development	Sridhara, S.: eds.	9788189422998	2009
47	Women in Sustainable Agriculture	C.Satapathy & Sabita Mishra	9789383305056	2014
48	Youth in Agriculture and Rural Development	K. Narayana Gowda:Ed.	9789381450758	2013

188 E-2 Rs. 1500/-